Emerging Materials for Energy and Sensing Applications

Editors

Kuldeep Singh Gour
CSIR-National Metallurgical Laboratory
Jamshedpur, India

Vidya Nand Singh
CSIR-National Physical Laboratory
New Delhi, India

CRC Press
Taylor & Francis Group
Boca Raton London New York

CRC Press is an imprint of the
Taylor & Francis Group, an **informa** business

Cover credit: Cover design prepared by the editors of the book using copyright free image.

First edition published 2025
by CRC Press
2385 NW Executive Center Drive, Suite 320, Boca Raton FL 33431

and by CRC Press
4 Park Square, Milton Park, Abingdon, Oxon, OX14 4RN

CRC Press is an imprint of Taylor & Francis Group, LLC

Library of Congress Cataloging-in-Publication Data (applied for)

ISBN: 978-1-032-64619-0 (hbk)
ISBN: 978-1-032-67367-7 (pbk)
ISBN: 978-1-032-67369-1 (ebk)

DOI: 10.1201/9781032673691

Typeset in Times New Roman
by Prime Publishing Services

Preface

This book is the outcome of combined efforts by well-recognized researchers in this research field. The proposed book explores the latest advancements in materials science for energy generation, storage, and sensing technology. It covers various topics, including nanomaterials, thin film solar cells, emerging materials-based solar cells, quantum dots, gas sensors, thermoelectric, photodetectors, etc. This book highlights emerging materials' applications, making it an invaluable resource for researchers, engineers, and students looking to stay up-to-date in this rapidly evolving field. Its focus on cutting-edge materials and their potential to develop energy harnessing and information gathering provides a deep insight into the exciting world of innovative materials. This book contains seven chapters that cover emerging materials and their application in energy and sensors. The book aims to address the needs of researchers in the areas of new materials for energy & sensor applications.

Chapter 1 explains electrochemical water splitting presents significant potential as an effective method to generate a portable and sustainable energy source with hydrogen fuel. The critical factor in moving closer to its integration into industries relies on the adoption of cost-effective electrocatalysts that exhibit high efficiency. A comprehensive overview of recent developments in earth-abundant Fe-based self-supported electrocatalysts for water splitting and alternative oxidation reactions for efficient hydrogen production is explained in this chapter. This chapter aims to foster progress in catalyst design and development and inspire future research in electrochemical water splitting for sustainable hydrogen production.

Chapter 2 explains the need for sustainable energy solutions; the field of thermoelectric materials has emerged as a pivotal area of research, aiming to harness waste heat and convert it into valuable electrical energy. The chapter gives an overview of magnesium-based Zintl phases, when engineered precisely with selected alloying elements, exhibit distinctive electronic and thermal properties that can be tuned for improved thermoelectric performance and unlock the potential of magnesium-based Zintl compounds for sustainable energy conversion.

Chapter 3 explains electrochemical water splitting, a pivotal technology for clean hydrogen production, hinges on developing efficient catalysts to expedite the challenging oxygen evolution reaction (OER) and hydrogen evolution reaction (HER). This chapter elucidates the fundamental concepts of defects in catalyst materials, encompassing oxygen vacancies, metal cation vacancies, and edge defects. Furthermore, it explores how these defects can facilitate charge transfer kinetics, modulate surface reactivity, and influence the overall electrocatalytic activity of materials. This chapter can serve as a comprehensive guide for researchers, engineers, and practitioners seeking to harness the influence of defects in electrochemical water splitting.

Chapter 4 explains two-dimensional (2D) materials like MoS_2, WS_2, WSe_2, and $MoSe_2$ and their application in broadband photodetection. This chapter gives insights into how photodetector devices with heterostructures drastically improve the photocurrent, detection range, and response time compared to only transition metal dichalcogenides (TMDs).

Chapter 5 explains metal oxide nanoparticles have potential uses in energy and the environment, including solar cells, fuel cells, hydrogen energy, and energy storage devices. This chapter overviews current scientific advancements in producing nanostructured metal oxides for energy conversion and storage devices, along with some important difficulties and prospects.

Chapter 6 explains the basics of gas sensing and synthesizing materials for sensor devices. This chapter discusses how ambient gases alter the surface and free charge carrier concentration and the factors affecting selenide semiconductivity. Furthermore, the most fascinating gas-sensitive materials are discussed, along with their unique characteristics.

Chapter 7 explains that the requirement for superior-quality gas sensors is becoming inevitable, especially in detecting hazardous gases that cause air pollution. Moreover, the non-invasive techniques of early-stage detection of various diseases by detecting VOCs in the exhaled breath of affected patients are gaining popularity, and ultra-sensitive gas sensors are also required. This chapter describes brief historical development from chemiresistive metal oxide-based gas sensors (e.g. ZnO, SnO_2, In_2O_3, etc.) along with several drawbacks to the strategies to develop modern 2D material-based room temperature operated gas sensors (e.g. graphene, MoS_2, WS_2, SnS_2, hBN, etc.) with superior sensing properties, shift of the operating principle from chemisorption (for oxide based sensor) to physisorption (for 2D material based sensor) illustrated. These emerging 2D materials can be expected to revolutionize gas sensing technology.

Kuldeep Singh Gour
Vidya Nand Singh

Contents

CHAPTER
1

Free-standing Nanostructured Iron-based Electrocatalysts for Water Splitting and Selective Oxidation Processes

Mayur A. Gaikwad* **and Dhanaji B. Malavekar**
Department of Materials Science and Engineering and Optoelectronics Convergence Research Center, Chonnam National University, Gwangju 61186, Republic of Korea

1. Importance of Catalysts in Electrocatalysis

Utilizing electrocatalytic water splitting is a promising approach towards transforming sustainable electrical energy into valuable H_2 fuel, which is in high demand [1]. The effectiveness of the process of electrocatalysis depends heavily on the catalytic activity of the electrocatalysts [2]. Therefore, it is essential to improve the basic performance of the electrocatalysts to optimize the energy conversion efficiency, here the water-splitting reaction. Furthermore, the expense of electrocatalysts and the techniques used to prepare them can have a substantial influence on the total cost of H_2 production [3]. The process of electrolytic water splitting, requires overcoming a considerable kinetic barrier. When it comes to decreasing the kinetic barrier, electrocatalysts are an extremely important factor. Several essential characteristics for activity, stability, and efficiency are used to evaluate the performance of an electrocatalyst for electrocatalytic water splitting. This evaluation is based on parameters like overpotential, Tafel slope and exchange current density that define the activity [4].

2. Significance of Water Splitting and Selective Oxidation Reactions

Water splitting and selective oxidation reactions are significant in the context of

*Corresponding author: mayur00gaikwad@gmail.com

energy conversion, storage and sustainable chemical processes. H_2 is considered a clean and versatile energy source and its combustion or use in fuel cells generates water as the only byproduct, making it a promising alternative to fossil fuels. The water splitting system can be integrated into renewable energy conversion systems to convert excess electricity generated during peak times into H_2 fuel [5]. Selective oxidation reactions are crucial in green chemistry as they allow for the targeted conversion of specific compounds without producing unwanted byproducts. This approach via other oxidation reactions, not only reduces energy consumption but also generates value-added products alongside H_2 (Figure 1). This is important for minimizing byproduct and waste production and the environmental impact of chemical processes [6].

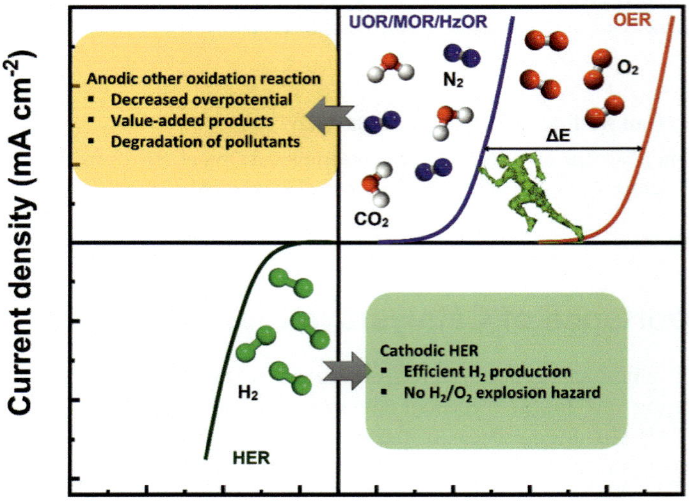

Figure 1. Differentiation of water electrolysis with the sluggish OER and with other oxidation reactions (UOR, MOR, and HzOR) coupled with HER.

3. A Brief Overview of State-of-art HER and OER Electrocatalysts and the Need for Self-supported Low-cost Electrocatalysts

Among various electrocatalysts, some noble metals are regarded as state-of-art electrocatalysts such as platinum (Pt), ruthenium (Ru), palladium (Pd) and iridium (Ir), which provide outstanding electrocatalytic activity for HER and OER [7]. Facilitated by their appropriate H_2 binding energies, Pt-based catalysts have usually acted as the benchmark for HER and they exhibit relatively high HER activity. However, the exchange current density of Pt-based catalysts in alkaline media is about 2-3 orders of magnitude lower than that in acidic media owing to their poor water dissociation ability [8]. With a similar H_2 bond energy

(~65 kcal mol^{-1}) as Pt, Ru is also and nearly ideal HER catalyst; further, the price of Ru is one-fourth that of Pt, which makes it a good alternative for use in HER. In addition, Ru even exhibits a stronger water dissociation ability than Pt metal. Similarly, Pd and Ir have emerged as a significant metal concerning electrochemical water splitting applications owing to their lower cost than Pt electrodes [9]. Pd is also a worthwhile H_2 host and serves as a model system of H absorption and desorption. In this regard, Pd-based materials are a promising alternative for use in water reduction. As proven in earlier works, H_2 binding energy can be the single description of HER activity both in acidic and alkaline media. However, the expensiveness and scarcity of these metals make them less favorable for scalable applications. Therefore, transition metal-based compounds like hydroxides, oxides, sulfides, phosphides, etc. have been employed for electrocatalytic reactions [10]. However, there are several challenges like low catalytic activity, sluggish kinetics of OER, poor low durability, and scalability associated with transition metal-based electrocatalysts used in water splitting systems [11]. These obstacles present an opportunity for further research into advancement in transition metal-based electrocatalysts and innovative device architectures that can achieve efficient and stable H_2 production. Considerable research efforts have been dedicated to the development of efficient Iron (Fe)-based electrocatalysts for both the HER and the OER. In recent years, there has been extensive exploration and utilization of various Fe-based compounds, including oxyhydroxides, hydroxides, oxides, sulfides, carbides, nitrides, phosphates, and phosphides, as well as their composites as electrocatalysts in electrochemical water splitting systems [12]. Researchers have been investigating these diverse materials and their combinations to enhance the efficiency and effectiveness of electrocatalysis in water-splitting processes. Additionally, self-supported electrodes offer several key advantages compared to conventional electrodes. Firstly, they allow effective anchoring and dispersion of active catalysts on the substrate, enabling higher loading amounts and a greater number of active sites. Secondly, they eliminate the need for polymeric binders and conductive additives, resulting in increased exposed catalytically active sites. Thirdly, self-supported electrodes can be engineered with nanostructured configurations, such as one-dimensional (1D), two-dimensional (2D), or three-dimensional (3D) structures, leading to enhanced electroactive surface area, efficient charge transport and improved reaction kinetics. Lastly, the strong adhesion and seamless integration between electrocatalysts and conducting substrates prevent the electrocatalyst detachment during continuous operation of water splitting, even at high current densities.

This book chapter provides a thorough examination of the latest developments in free-standing nanostructured Fe-based electrocatalysts for electrochemical water splitting and specific oxidation reactions, including urea oxidation reaction (UOR), methanol oxidation reaction (MOR), and hydrazine oxidation reaction (HzOR). Subsequently, we discuss an extensive array of free-standing Fe-based electrocatalysts, including oxyhydroxides, hydroxides, oxides, sulfides and phosphides, along with their composites. We investigate the methods of designing, creating and modulating the structure, introducing dopants, and constructing heterostructures of these electrocatalysts to enhance the rate of chemical reactions and he overall efficiency of the process of water splitting. In addition, we focus on the challenges that need to be overcome to enhance the effectiveness and longevity of Fe-based electrocatalysts for practical use. In conclusion, the book chapter addresses the obstacles and prospects in this field, incorporating the contributions of researchers from materials science, chemistry, electrochemistry and related fields.

4. Catalyst Support Materials: Significance and Impact on Catalytic Activity

The choice of suitable support material is a crucial factor in achieving improved catalytic efficiency of the free-standing electrocatalysts for water splitting and selective alternative oxidation. Exemplary support material should exhibit several crucial characteristics, including high mechanical strength, exceptional electronic conductivity, resistance to corrosion, a substantial surface area and a strong binding force to the catalyst. Various types of supports, including metals (such as mesh, plate and foam) [13] and carbonaceous materials (such as carbon cloth, carbon fiber paper and graphene film) [14], have been extensively studied for the preparation of self-supporting electrocatalysts. The electrocatalysts are prepared directly on 3D structured conductive supports, such as metal or carbonaceous materials and have various advantages. They are advantageous because they do not include any additives, they have a large number of active sites, which implies that they can accommodate a greater quantity of electrocatalysts. The second benefit is that the electrocatalysts that are evenly dispersed on the supports make it easier for beneficial charge transport to occur. When it comes to surface engineering, the morphology and microstructure of these electrocatalysts can be simply modified to achieve the desired final result. Another factor that contributes to the electrocatalyst's long-term stability is the anchoring structure or connection that exists between the electrocatalyst and the supports. Presently, there has been significant research conducted on 3D structured metal foams to provide support to electrocatalysts. Despite the outstanding conductivity displayed by metal foam-based supports, their mechanical strength and chemical stability are insufficient to ensure long-term stability in actual applications involving huge current density and high operation temperature. Gu et al. [15] conducted a comparative investigation on NiFe-LDH electrocatalysts that were built rationally and used different supports. The electrochemical results demonstrated a significant enhancement in the activity of NiFe-LDH when utilizing Co metal support, resulting in a tenfold increase compared to conventional inert conductive supports like FTO, ITO and glassy carbon. The author observed superior electrocatalytic capabilities for electrocatalysts deposited on metal-based supports compared to those on TCO. During device operation, the interface between the electrocatalyst and metal supports can easily generate interfacial compounds, such as oxide and/or hydroxide. This might result in the loss of the device's performance. The selection of a particular synthesis technique and support system for creating self-supported electrocatalysts depends on various criteria, including the desired characteristics of the electrocatalyst and the planned use of the electrocatalytic reaction. In electrochemical deposition methods, metal-based substrates are favored for their superior electron conductivity. Conversely, in hydro/solvothermal synthesis, both metal foams and carbon-based substrates are primarily utilized for their extensive surface area.

5. Progresses in Fe-based Electrocatalyst Design and Performance

The process of electrochemical water splitting is a chemical procedure that involves the separation of water into its fundamental components, namely H_2, and Oxygen, with the assistance of external energy. Two half-reactions are involved in this process: the first is HER at the cathode, and the second is OER at the anode. Based on the pH of the electrolyte, the electrical properties of the electrocatalyst and the features of the electrode surfaces, these reactions proceed along a variety of distinct reaction routes. Both the cathode and the anode are capable of acting as catalysts for the water reduction and oxidation reactions concurrently. To overcome the activation energy barriers and the slow reaction kinetics that are taking place at each electrode, it is necessary to have an additional voltage that is greater than the typical 1.23 V vs. RHE. This additional voltage is referred to as the "overpotential". Recent years have seen significant advancements in water electrolysis, with the primary focus being on addressing these issues. This has been accomplished by substituting the slow OER with alternate oxidation reactions that offer greater benefits, such as urea oxidation reaction (UOR), methanol oxidation reaction (MOR) and hydrazine oxidation reaction (HzOR). When compared to OER, these alternative reactions typically display theoretical overpotentials that are lower (Figure 1).

In recent times, it has been discovered that compounds based on Fe have demonstrated remarkable outcomes by providing compromises in terms of lowered overpotentials and long-term stabilities in high current densities [16]. They have been investigated extensively as oxides, including bi- and tri-metallic oxides, with varying stoichiometry, to achieve effective activity with very low overpotentials, particularly in OER [17]. Furthermore, Fe hydroxides have achieved a breakthrough in the form of mono-metallic hydroxides and layered double hydroxides (LDH), both of which have increased their performance in water splitting [18]. Additionally, current developments on sulfides and selenides are exhibiting increased OER and HER activities [19]. This is in addition to the previous point. Regarding alternative electrocatalysts, metal phosphides are the most important ones since they demonstrate significant enhancements in both OER and HER under a variety of pH circumstances [20]. The energy input is significantly lowered with these sorts of phosphide-based catalysts, which offer high-rate performance in the entire water splitting process. In this regard, Fe-based phosphides are delivering high-rate performances. Recently, the presence of phosphides in OER has been investigated for the developments that have been made and it has the potential to bring about "synergistic enhancements" to the electrocatalytic activities. When it comes to HER, the presence of phosphides, sulfides and selenides that possess enough amount of electronegativity can successfully collect H_2 ions from the acidic electrolyte. During the HER and OER mechanistic phases, the presence of metal sulfides/selenides/phosphides demonstrated increased charge transfer, as well as a high electrical conductivity in comparison to their oxide equivalents [21]. Therefore, it is of great interest to obtain a visual representation of the development and significance of the Fe-based electrocatalyst with electrocatalytic water splitting, which has shown such significant advancements in terms of alternate oxidation reactions. A summary of the self-supported Fe-based electrocatalysts used as bifunctional electrocatalysts and for other oxidation reactions is also provided in Tables 1 and 2.

6. Advances in Electrocatalysts Stability

Typically, the stability assessment of electrocatalysts involves the use of cyclic voltammetry (CV), chronopotentiometry (CP) and chronoamperometry (CA) methods. While there has been extensive research and notable advancements in the study of the catalytic activity of electrocatalysts, their stability has received comparatively less attention. Although many of above discussed electrocatalysts demonstrate stability for periods ranging from tens to hundreds of hours under electrochemical conditions, such short-term stability falls short of meeting the demands for practical applications. Practically, the effective utilization of the electrocatalysts in water electrolysis necessitates robust stability to function efficiently under higher current density for commercial applications [13c]. Nevertheless, a significant challenge arises as these Fe-based metal oxides undergo considerable corrosion during the OER study, imposing a substantial constraint on the stability of electrocatalysts. Understanding catalyst deactivation holds paramount importance in elucidating the stability of electrocatalysts and it plays a crucial role in the rational design of electrocatalysts with enhanced stability [46]. Catalyst dissolution manifests through both chemical and electrochemical processes. Chemically, dissolution arises from the catalyst interacting with electrolyte components or impurities. In contrast, electrochemical dissolution occurs when the catalyst undergoes oxidation or reduction under an applied potential. Deactivation of electrocatalysts also can be initiated by alterations in the crystal structure and morphology of the catalyst. According to the Pourbaix diagram, it is proposed that the accelerated conversion of FeOOH into soluble

Table 1. Summary of Fe-based self-supported bifunctional electrocatalysts

Material system	Support	Electrolyte	Overpotential (mV) at j (mA cm^{-2})		Tafel slope (mV dec^{-1})		Overall water splitting at j (mA cm^{-2})		Ref.
			HER	OER	HER	OER	Cell voltage (V)	Stability Time (h)	
CoFe	NF	1M KOH	110 at 10	220 at 10	35	40	1.64 at 10	50 at 10	[22]
F0.25C1CH	NF	1M KOH	77 at 10	228 at 10	99	42	1.45 at 10	100 at 100	[23]
Cu_2O@Fe_2O_3	CC	1M KOH	188 at 10	296 at 10	59	66	1.67 at 10	20 at 10	[24]
Na-Fe-Pt	NF	1M KOH	31 at 10	261 at 10	35.98	39.68	1.56 at 10	12 at 10	[25]
NiFeLDH/Ni$(OH)_2$	NF	1M KOH	180 at 20	370 at 700	112	126	1.6 at 10	20 at 500	[18b]
Ni-FeO_x/$FeNi_3$	NF	1M KOH	71 at 150	269 at 50	44.6	78.6	1.58 at 50	100 at 30	[26]
Ni–Fe–P/Ni	CS	1M KOH	66 at 10	198 at 10	60	60	1.50 at 10	10 at 100	[27]
NiFeP-MoO_2	NF	1M KOH	56 at 10	143 at 10	80.5	29	1.41 at 10	550 at 300	[28]
MnO_x/NiFeP	NF	1M KOH	255 at 500	296 at 500	41.8	59.8	1.68 at 10	120 at 500	[29]
Ni90Fe10-PC3000-2A	Ni/Au (AAO)	1M KOH	100 at 10	236 at 10	78	45	1.58 at 10	24 at 400	[30]
FeIr	NF	1M KOH	25.6 at 10	220 at 10	50.46	64.78	1.51 at 10	100 at 500	[31]
IrNi-$FeNi_3$	NF	1M KOH	288.8 at 1000	330 at 1000	66.95	36.01	1.47 at 10	100 at 500	[32]
NiFeMo IOS	NF	1M KOH	249 at 500	293 at 500	112	170	1.47 at 10	50 at 500	[33]

Table 2. Summary of Fe-based self-supported electrocatalysts used for selective UOR, MOR, and HzOR

Material system	Support	Electrolyte	Overpotential (η) (mV) at j (mA cm^{-2})			Tafel slope (mV dec^{-1})			Overall water splitting at j (mA cm^{-2})		Ref.
			HER	OER	Selective oxidation reaction potential (V) at j (mA cm^{-2})	HER	OER	Selective oxidation reaction	Cell voltage (V)	Stability Time (h)	
NiFeCo LDHs	NF	1M KOH + 0.33 M Urea	108 at 10	210 at 10	0.280 at 10	64	39	31	1.49 at 10	50 at 10	[34]
NiFeOOH	NF	1M KOH + 0.33 M Urea	-	230 at 50	1.38 at 50	-	44	57	-	75 at 50	[35]
FeCo-LDH	NF	1M KOH + 0.33 M Urea	-	228 at 50	1.35 at 50	-	56.6	28.8	1.62 at 50	24 at 50	[36]
FeOOH@ Co_3O_4-240	NF	1M KOH + 0.33 M Urea	174 at 10	228 at 10	1.33 at 10	57	71	29	1.43 at 10	12 at 50	[37]
Fe_2O_3 and FeP	ECP	1M KOH + 100 mM Hydrazine	138 at 10	330 at 10	0.70 at 10	63.9	179.2	-	0.93 at 10	20 at 1V	[38]
FeP-NS	NF	1M KOH + 0.01 M methanol	-	220 at 100	-	-	51	-	-	85 at 10	[39]
FeMn-PS	NF	1M KOH + 0.33 M Urea	-	237 at 100	1.35 at 100	-	59	19	1.36 at 100	12 at 100	[40]

$Fe_2P@Ni_xP$	NF	1M KOH + 0.5 M Urea	276 at 100	254 at 100	1.347 at 100	91	-	30	1.53 at 100	30 at 100	[41]
Fe-($Ni_{12}P_5$/ Ni_3P)	NiFe foam	1M KOH + 0.5 M Urea	-	300 at 100	1.35 at 100	-	28.2	82.8	-	105 at 10	[42]
P-NiFeP/Ni	Ni	1M KOH + 0.1 M N_2H_4	17.9 at 10	-	0.077 at 10	63	-	83.6	0.16 at 10	55 at 100	[43]
Fe-Ni_3S_2/ NF-2	NF	1M KOH + 0.33 M Urea + 1 M Hydrazine	-	360 at 100	1.39 at 100 0.36 at 100	-	-	32	1.85 at 50	80 at 10	[44]
Fe-Ni_3S_2	NF	1M KOH + 0.33 M Urea	254 at 100	290 at 100	198	-	-	-	1.57 at 100	500 at 500	[45]

FeO_4^{2-} occurs more rapidly at higher dissolution rates corresponding to more anodic applied potentials (Figure 2a). As shown in Figure 2b-c, numerous studies indicate that upon conducting post-stability analyses of electrocatalysts, the genuine active sites are identified as their respective oxyhydroxides [4a, 26, 47]. Similarly, at higher potentials or current densities, the formation of an insulating layer at the interface between the catalyst and the substrate electrode may occur. This insulating layer can impede electron transmission, consequently causing deactivation.

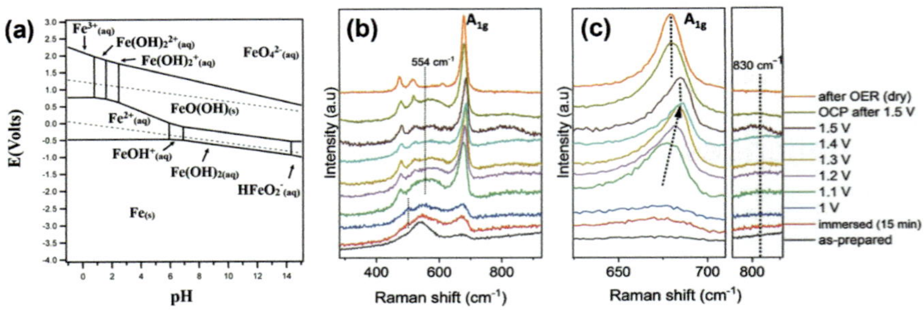

Figure 2. (a) E-pH diagram for Fe species, (b) In situ Raman spectra of the Co/Fe 32-red sample measured in a 1 M KOH electrolyte within a 0.1 V potential step. (c) Magnification of the A1g band (left) and active oxygen species (right).

Typically, electrocatalysts are deposited onto a substrate and detachment may arise when the interaction between the electrocatalysts and the substrate weakens. This detachment can be attributed to factors such as corrosion or gas evolution. Consequently, the significance of self-supported electrocatalysts becomes evident, addressing potential challenges associated with detachment and enhancing overall stability [44]. Furthermore, if evolved gas bubbles formed cannot detach promptly, the catalyst's surface may be obstructed by gas bubbles, creating a barrier between the catalyst and the electrolyte. This hinders ionic transportation, resulting in deactivation. This concern is particularly pronounced for electrocatalysts exhibiting high activity at higher potential or current density. The stability of electrocatalysts is influenced by various factors, including the catalyst's intrinsic properties such as crystalline and electronic structures and morphology. Doping strategy as well as the nature of the substrate electrode, also play crucial roles [3b]. OER conditions like applied potential or current density, electrolyte composition, and temperature, further contribute to electrocatalyst stability. Strategic adjustments to these factors can be involved in enhancing the overall stability of electrocatalysts, highlighting the importance of a comprehensive approach in catalyst design and optimization.

Enhancing the stability of electrocatalysts involves strategic measures to counter deactivation factors. Given that dissolution is a major contributor to electrocatalyst deactivation, establishing a protective layer demonstrates heightened resistance against dissolution during catalytic activity, thereby promoting stability. To address structural changes leading to deactivation at high anodic potentials, cathodic treatment serves as a rejuvenating approach by facilitating surface site refreshment through reduction, dissolution and re-deposition, ultimately contributing to reactivation [48]. Additionally, preventing surface blockage caused by gas bubble formation is pivotal for sustained catalytic performance. Specific electrocatalyst morphologies offer a means to circumvent the bubble-induced blockage. The implementation of a super aerophobic nanoarray emerges as a strategic approach, mitigating gas bubble adhesion, expanding bubble contact angles and facilitating the rapid release of small gas bubbles [49]. This approach is recognized as a crucial strategy for achieving and maintaining stable OER performance.

7. Challenges and Opportunities in Electrocatalyst Design

The global pursuit of efficient and sustainable energy sources has heightened interest in water electrolysis for green H_2 fuel production. Despite significant advancements, the field of electrocatalyst design faces challenges that demand innovative solutions. Addressing these challenges is essential for unlocking the full potential of catalytic systems and ensuring their stability, scalability and integration into diverse applications.

7.1 Challenges

7.1.1 Stability and Scalability

The stability of electrocatalysts over prolonged use and their scalability for large-scale applications remain key challenges. Ensuring long-term stability and cost-effective scalability are critical for the widespread adoption of water electrolysis [50]. Water electrolysis functions under challenging conditions, including higher temperatures, corrosive electrolytes and dynamic potential cycling. It is vital to engineer electrocatalysts that can sustain their activity over extended periods without experiencing degradation. The potential deterioration of electrocatalysts, attributed to factors such as electrolyte corrosion, metal leaching, or structural alterations, can result in reduced performance and increased operational expenses [51]. Consequently, the design of electrocatalysts must select materials with stability and durability under these rigorous operating conditions.

7.1.2 Selectivity and Energy Efficiency

Ensuring the selectivity of electrocatalysts for targeted reactions is crucial to reducing side reactions that could compromise the overall efficiency of water electrolysis. In specific instances, such as alkaline electrolyzers, the HER may put up with the undesired H_2 oxidation reaction (HOR), leading to a reduction in overall efficiency [6b]. Consequently, it becomes necessary to design electrocatalysts that not only promote the desired reactions selectively but also effectively suppress the occurrence of unwanted reactions by reducing energy consumption.

7.1.3 Precious Metal Dependency

The economic viability of large-scale water electrolysis is significantly influenced by the cost of catalyst materials. Current electrocatalysts often rely on precious metals like platinum and iridium, posing economic challenges due to their scarcity and high costs [52]. Reducing dependency on these precious metals without compromising performance is a significant hurdle. Additionally, catalyst synthesis methods should be scalable to meet the demands of industrial production.

7.1.4 Electrode Fabrication and Effective Mass Transport

Electrocatalysts are commonly deposited on conductive supports to improve stability and facilitate easy charge transfer. However, achieving a uniform distribution, establishing effective electronic coupling between the electrocatalyst and the supporting material and ensuring robust adhesion can pose challenges [53]. Furthermore, issues such as electrode delamination, instability and reduced activity may arise due to inadequate electrode integration. Moreover, efficient mass transport of reactants and products to and from the electrocatalyst surface is essential for overall efficiency [27]. Overcoming limitations related to mass transport, especially in nanomaterial-based catalysts, is a significant challenge.

7.2 Opportunities

The field of catalyst design for water electrolysis is prevailing with opportunities that hold immense potential for advancing the efficiency and sustainability of this critical technology. This field offers a wide range of opportunities in catalysts design from tailoring active sites at the atomic and molecular levels to leveraging nanomaterials and exploring earth-abundant alternatives. Bifunctional and multifunctional electrocatalysts and advancements in computational and machine learning approaches also present exciting prospects. Additionally, the integration of water electrolysis with renewable energy systems and their industrial applications offers promising opportunities for alignment with existing infrastructure.

7.2.1 Active Site Engineering

The active sites of electrocatalysts serve as key locations where electrochemical reactions take place. Tailoring catalysts at the atomic and molecular levels offers opportunities to enhance activity, selectivity and stability [5]. Understanding and controlling active sites can open new avenues for optimizing electrocatalyst performance.

7.2.2 Nanomaterials and Nanostructuring

Exploiting the unique properties of nanomaterials, such as nanoparticles, nanowires and nanosheets, provides a promising platform for electrocatalyst innovation. This approach enhances surface areas, mass transport efficiency nd electronic structures, contributing to improved catalytic activity [54]. Investigating innovative nanostructuring techniques provides researchers with the means to customize catalysts for specific reactions, thereby optimizing their overall performance.

7.2.3 Earth-abundant Bifunctional and Multifunctional Electrocatalysts

Exploring electrocatalysts based on earth-abundant materials, such as iron and nickel, presents an opportunity to an opportunity to replace precious metals like platinum and iridium. The design of catalysts with the capability to promote multiple reactions simultaneously, including both OER and HER, streamlines the electrolysis process. This incorporation of multifunctional capabilities not only optimizes performance by promoting selectivity but also effectively suppresses side reactions [55].

7.2.4 Operando Spectroscopy and In Situ Characterization

To establish foundational design principles for highly efficient electrocatalysts, a profound comprehension of the catalytic mechanism and the role played by iron (Fe) is indispensable. The complexity introduced by the presence of multiple metals in these electrocatalysts adds an additional layer of complexity to the interaction between their atomic and electronic structures, posing challenges in understanding these materials at the molecular level. Overcoming these challenges necessitates the implementation of advanced in-situ or operando techniques, providing real-time observation and insights into reactive sites. Utilizing spectroscopic techniques such as X-ray absorption (XAS), X-ray photoelectron spectroscopy (XPS), Raman spectroscopy, Fourier-transform infrared spectroscopy (FTIR), nuclear resonant inelastic X-ray scattering, Mössbauer spectroscopy and electrochemical impedance spectroscopy (EIS), alongside microscopic techniques like atomic force microscopy (AFM) and transmission electron microscopy (TEM), becomes crucial [56]. The application of these techniques is instrumental in unveiling the nature of reactive sites, facilitating the design, and enabling the fabrication of self-supported electrocatalysts for water splitting characterized by exceptional activity and stability.

References

1. L. Kumar, B. Antil, A. Kumar, M.R. Das, O. López-Estrada, S. Siahrostami, S. Deka, Experimental and computational insights into the overall water splitting reaction by

the Fe–Co–Ni–P electrocatalyst. ACS Applied Materials & Interfaces, 15 (2023) 54446-54457.
2a. L. Jing, R. Zhu, D.L. Phillips, J.C. Yu, Effective prevention of charge trapping in graphitic carbon nitride with nanosized red phosphorus modification for superior photo (electro) catalysis. Advanced Functional Materials, 27 (2017) 1703484.
2b. S. Anwar, F. Khan, Y. Zhang, A. Djire, Recent development in electrocatalysts for hydrogen production through water electrolysis. International Journal of Hydrogen Energy, 46 (2021) 32284-32317.
3a. P. Su, W. Fu, X. Du, J. Guo, M. Zhou, Cost-effective degradation of pollutants by in-situ electrocatalytic process on Fe@BN-C bifunctional cathode: Formation of $1O_2$ with high selectivity under nanoconfinement. Chemical Engineering Journal, 452 (2023) 139693.
3b. M.-S. Balogun, W. Qiu, Y. Huang, H. Yang, R. Xu, W. Zhao, G.-R. Li, H. Ji, Y. Tong, Cost-effective alkaline water electrolysis based on nitrogen- and phosphorus-doped self-supportive electrocatalysts. Advanced Materials, 29 (2017) 1702095.
4a. D. Yao, L. Gu, B. Zuo, S. Weng, S. Deng, W. Hao, A strategy for preparing high-efficiency and economical catalytic electrodes toward overall water splitting. Nanoscale, 13 (2021) 10624-10648.
4b. J. Hou, Y. Wu, B. Zhang, S. Cao, Z. Li, L. Sun, rational design of nanoarray architectures for electrocatalytic water splitting. Advanced Functional Materials, 29 (2019) 1808367.
5. J. Huang, Y. Jiang, T. An, M. Cao, Increasing the active sites and intrinsic activity of transition metal chalcogenide electrocatalysts for enhanced water splitting. Journal of Materials Chemistry A, 8 (2020) 25465-25498.
6a. D. Li, Y. Xing, R. Yang, T. Wen, D. Jiang, W. Shi, S. Yuan, Holey cobalt–iron nitride nanosheet arrays as high-performance bifunctional electrocatalysts for overall water splitting. ACS Applied Materials & Interfaces, 12 (2020) 29253-29263.
6b. K. Xiang, D. Wu, X. Deng, M. Li, S. Chen, P. Hao, X. Guo, J.-L. Luo, X.-Z. Fu, Boosting H_2 generation coupled with selective oxidation of methanol into value-added chemical over cobalt hydroxide@hydroxysulfide nanosheets electrocatalysts. Advanced Functional Materials, 30 (2020) 1909610.
7a. R. Samanta, P. Panda, R. Mishra, S. Barman, IrO_2-modified RuO_2 nanowires/nitrogen-doped carbon composite for effective overall water splitting in all pH. Energy & Fuels, 36 (2022) 1015-1026.
7b. P. Li, X. Duan, Y. Kuang, X. Sun, Iridium in tungsten trioxide matrix as an efficient Bi-functional electrocatalyst for overall water splitting in acidic media. Small, 17 (2021) 2102078.
8a. X. Zheng, P. Cui, Y. Qian, G. Zhao, X. Zheng, X. Xu, Z. Cheng, Y. Liu, S.X. Dou, W. Sun, Multifunctional active-center-transferable platinum/lithium cobalt oxide heterostructured electrocatalysts towards superior water splitting. Angewandte Chemie International Edition, 59 (2020), 14533-14540.
8b. X. Yu, J. Guo, B. Li, J. Xu, P. Gao, K.S. Hui, K.N. Hui, H. Shao, Sub-nanometer Pt clusters on defective NiFe LDH nanosheets as trifunctional electrocatalysts for water splitting and rechargeable hybrid sodium–air batteries. ACS Applied Materials & Interfaces, 13 (2021) 26891-26903.
9a. F. Lv, W. Zhang, W. Yang, J. Feng, K. Wang, J. Zhou, P. Zhou, S. Guo, Ir-based alloy nanoflowers with optimized hydrogen binding energy as bifunctional electrocatalysts for overall water splitting. Small Methods, 4 (2020) 1900129.

9b. Z. Chen, X. Duan, W. Wei, S. Wang, B.-J. Ni, Iridium-based nanomaterials for electrochemical water splitting. Nano Energy, 78 (2020) 105270.
10a. M. Li, X. Liu, X. Hu, Fabrication of core–sheath NiCoP@FePx nanoarrays for efficient electrocatalytic hydrogen evolution. ACS Sustainable Chemistry & Engineering, 6 (2018) 8847-8855.
10b. Z. Kou, K. Wang, Z. Liu, L. Zeng, Z. Li, B. Yang, L. Lei, C. Yuan, Y. Hou, Recent advances in manifold exfoliated synthesis of two-dimensional non-precious metal-based nanosheet electrocatalysts for water splitting. Small Structures, 3 (2022) 2100153.
11a. S. Roy, D. Bagchi, L. Dheer, S.C. Sarma, V. Rajaji, C. Narayana, U.V. Waghmare, S.C. Peter, Mechanistic insights into the promotional effect of Ni substitution in non-noble metal carbides for highly enhanced water splitting. Applied Catalysis B: Environmental, 298 (2021) 120560.
11b. H. Xu, K. Ye, K. Zhu, Y. Gao, J. Yin, J. Yan, G. Wang, D. Cao, Transforming carnation-shaped MOF-Ni to Ni-Fe prussian blue analogue derived efficient bifunctional electrocatalyst for urea electrolysis. ACS Sustainable Chemistry & Engineering, 8 (2020) 16037-16045.
12a. K. Srinivas, Y. Lu, Y. Chen, W. Zhang, D. Yang, $FeNi_3$–Fe_3O_4 heterogeneous nanoparticles anchored on 2D MOF nanosheets/1D CNT matrix as highly efficient bifunctional electrocatalysts for water splitting. ACS Sustainable Chemistry & Engineering, 8 (2020) 3820-3831.
12b. X. Cao, T. Wang, L. Jiao, Transition-metal (Fe, Co, and Ni)-based nanofiber electrocatalysts for water splitting. Advanced Fiber Materials, 3 (2021) 210-228.
12c. M. Yang, Y.-N. Zhou, Y.-N. Cao, Z. Tong, B. Dong, Y.-M. Chai, Advances and challenges of Fe-MOFs based materials as electrocatalysts for water splitting. Applied Materials Today, 20 (2020) 100692.
12d. Z. Wu, Y. Zhao, H. Wu, Y. Gao, Z. Chen, W. Jin, J. Wang, T. Ma, L. Wang, Corrosion engineering on iron foam toward efficiently electrocatalytic overall water splitting powered by sustainable energy. Advanced Functional Materials, 31 (2021).
13a. R.A. Marquez-Montes, K. Kawashima, Y.J. Son, J.A. Weeks, H.H. Sun, H. Celio, V.H. Ramos-Sánchez, C.B. Mullins, Mass transport-enhanced electrodeposition of Ni–S–P–O films on nickel foam for electrochemical water splitting. Journal of Materials Chemistry A, 9 (2021) 7736-7749.
13b. Y. Peng, K. Jiang, W. Hill, Z. Lu, H. Yao, H. Wang, Large-scale, low-cost, and high-efficiency water-splitting system for clean H_2 generation. ACS Applied Materials & Interfaces, 11 (2019) 3971-3977.
13c. T.E. Seuferling, T.R. Larson, J.M. Barforoush, K.C. Leonard, Carbonate-derived multi-metal catalysts for electrochemical water-splitting at high current densities. ACS Sustainable Chemistry & Engineering, 9 (2021) 16678-16686.
14a. M. Hassan, M.M. Baig, S. Yousaf, M. Faheem, A. Hussain, B. Niaz, F.A. Khan, S. Aziz, Y. Ahmed, K.H. Shah, S.G. Lee, Efficient water splitting catalyst: Low-temperature selenization of Co and Ni hydroxide nanosheets on carbon cloth for enhanced electro-catalytic activity. Diamond and Related Materials, 139 (2023) 110298.
14b. Z. Wang, C. Wei, X. Zhu, X. Wang, J. He, Y. Zhao, Dual functions of CO_2 molecular activation and 4f levels as electron transport bridges in erbium single atom composite photocatalysts therefore enhancing visible-light photoactivities. Journal of Materials Chemistry A, 9 (2021) 10150-10158.

14c. D. Yang, W. Hou, Y. Lu, W. Zhang, Y. Chen, Cobalt phosphide nanoparticles supported within network of N-doped carbon nanotubes as a multifunctional and scalable electrocatalyst for water splitting. Journal of Energy Chemistry, 52 (2021) 130-138.

15. H. Gu, G. Shi, H.-C. Chen, S. Xie, Y. Li, H. Tong, C. Yang, C. Zhu, J.T. Mefford, H. Xia, W.C. Chueh, H.M. Chen, L. Zhang, Strong catalyst–support interactions in electrochemical oxygen evolution on Ni–Fe layered double hydroxide. ACS Energy Letters, 5 (2020) 3185-3194.

16. L. Yaqoob, T. Noor, N. Iqbal, H. Nasir, N. Zaman, K. Talha, Electrochemical synergies of Fe–Ni bimetallic MOF CNTs catalyst for OER in water splitting. Journal of Alloys and Compounds, 850 (2021) 156583.

17. Y. Qiu, Z. Liu, Q. Yang, X. Zhang, J. Liu, M. Liu, T. Bi, X. Ji, Atmospheric-temperature chain reaction towards ultrathin non-crystal-phase construction for highly efficient water splitting. Chemistry – A European Journal, 28 (2022) e202200683.

18a. S. Ibraheem, G. Yasin, A. Kumar, M.A. Mushtaq, S. Ibrahim, R. Iqbal, M. Tabish, S. Ali, A. Saad, Iron-cation-coordinated cobalt-bridged-selenides nanorods for highly efficient photo/electrochemical water splitting. Applied Catalysis B: Environmental, 304 (2022) 120987.

18b. N.S. Gultom, H. Abdullah, C.-N. Hsu, D.-H. Kuo, Activating nickel iron layer double hydroxide for alkaline hydrogen evolution reaction and overall water splitting by electrodepositing nickel hydroxide. Chemical Engineering Journal, 419 (2021).

19a. G. Yilmaz, C.F. Tan, Y.-F. Lim, G.W. Ho, Pseudomorphic Transformation of interpenetrated Prussian blue analogs into defective nickel iron selenides for enhanced electrochemical and photo-electrochemical water splitting. Advanced Energy Materials, 9 (2019) 1802983.

19b. L. Huang, H. Wu, H. Liu, Y. Zhang, Phosphorous doped cobalt-iron sulfide/carbon nanotube as active and robust electrocatalysts for water splitting. Electrochimica Acta, 318 (2019) 892-900.

19c. Z. Jing, Q. Zhao, D. Zheng, L. Sun, J. Geng, Q. Zhou, J. Lin, Nickel-doped pyrrhotite iron sulfide nanosheets as a highly efficient electrocatalyst for water splitting. Journal of Materials Chemistry A, 8 (2020) 20323-20330.

20a. J. Rongé, T. Dobbelaere, L. Henderick, M.M. Minjauw, S.P. Sree, J. Dendooven, J.A. Martens, C. Detavernier, Bifunctional earth-abundant phosphate/phosphide catalysts prepared via atomic layer deposition for electrocatalytic water splitting. Nanoscale Advances, 1 (2019) 4166-4172.

20b. P.T. Babar, A.C. Lokhande, H.J. Shim, M.G. Gang, B.S. Pawar, S.M. Pawar, J.H. Kim, SILAR deposited iron phosphate as a bifunctional electrocatalyst for efficient water splitting. Journal of Colloid and Interface Science, 534 (2019) 350-356.

21a. J. Yin, J. Jin, H. Lin, Z. Yin, J. Li, M. Lu, L. Guo, P. Xi, Y. Tang, C.-H. Yan, Optimized metal chalcogenides for boosting water splitting. Advanced Science, 7 (2020) 1903070.

21b. A.P. Tiwari, K. Kim, S. Jeon, Improving intrinsic electrocatalytic activity of layered transition metal chalcogenides as electrocatalysts for water splitting. Current Opinion in Electrochemistry, 34 (2022) 100982.

22. P. Babar, A. Lokhande, H.H. Shin, B. Pawar, M.G. Gang, S. Pawar, J.H. Kim, Cobalt iron hydroxide as a precious metal-free bifunctional electrocatalyst for efficient overall water splitting. Small, 14 (2018).

23. L. Hui, Y. Xue, D. Jia, H. Yu, C. Zhang, Y. Li, Multifunctional single-crystallized

carbonate hydroxides as highly efficient electrocatalyst for full water splitting. Advanced Energy Materials, 8 (2018).
24. N. Zhang, Q. Zhang, C. Xu, Y. Li, J.-Y. Zhang, L. Wu, Y. Liu, Y.-Z. Fang, Z. Liu, Optional construction of Cu_2O@ Fe_2O_3@ CC architecture as a robust multifunctional photoelectronic catalyst for overall water splitting and CO_2 reduction. Chemical Engineering Journal, 426 (2021).
25. Y. Zhao, Y. Gao, Z. Chen, Z. Li, T. Ma, Z. Wu, L. Wang, Trifle Pt coupled with NiFe hydroxide synthesized via corrosion engineering to boost the cleavage of water molecule for alkaline water-splitting. Applied Catalysis B: Environmental, 297 (2021).
26. A. Qayum, X. Peng, J. Yuan, Y. Qu, J. Zhou, Z. Huang, H. Xia, Z. Liu, D.Q. Tan, P.K. Chu, F. Lu, L. Hu, Highly durable and efficient Ni-FeOx/$FeNi_3$ electrocatalysts synthesized by a facile in situ combustion-based method for overall water splitting with large current densities. ACS Applied Materials & Interfaces, 14 (2022) 27842-27853.
27. G. Barati Darband, M. Aliofkhazraei, S. Hyun, S. Shanmugam, Pulse electrodeposition of a superhydrophilic and binder-free Ni–Fe–P nanostructure as highly active and durable electrocatalyst for both hydrogen and oxygen evolution reactions. ACS Applied Materials & Interfaces (2020).
28. X. Wu, J. Li, Y. Li, Z. Wen, NiFeP-MoO_2 hybrid nanorods on nickel foam as high-activity and high-stability electrode for overall water splitting. Chemical Engineering Journal, 409 (2021).
29. P. Wang, Y. Luo, G. Zhang, M. Wu, Z. Chen, S. Sun, Z. Shi, MnOx-decorated nickel-iron phosphides nanosheets: Interface modifications for robust overall water splitting at ultra-high current densities. Small, 18 (2022) e2105803.
30. C.L. Huang, X.F. Chuah, C.T. Hsieh, S.Y. Lu, NiFe alloy nanotube arrays as highly efficient bifunctional electrocatalysts for overall water splitting at high current densities. ACS Applied Materials & Interfaces, 11 (2019) 24096-24106.
31. J. Chen, Y. Wang, G. Qian, T. Yu, Z. Wang, L. Luo, F. Shen, S. Yin, In situ growth of volcano-like FeIr alloy on nickel foam as efficient bifunctional catalyst for overall water splitting at high current density. Chemical Engineering Journal, 421 (2021).
32. Y. Wang, G. Qian, Q. Xu, H. Zhang, F. Shen, L. Luo, S. Yin, Industrially promising IrNi-FeNi3 hybrid nanosheets for overall water splitting catalysis at large current density. Applied Catalysis B: Environmental, 286 (2021).
33. C.-T. Hsieh, C.-L. Huang, Y.-A. Chen, S.-Y. Lu, NiFeMo alloy inverse-opals on Ni foam as outstanding bifunctional catalysts for electrolytic water splitting of ultra-low cell voltages at high current. Applied Catalysis B: Environmental, 267 (2020).
34. P. Babar, A. Lokhande, V. Karade, B. Pawar, M.G. Gang, S. Pawar, J.H. Kim, Bifunctional 2D electrocatalysts of transition metal hydroxide nanosheet arrays for water splitting and urea electrolysis. ACS Sustainable Chemistry & Engineering, 7 (2019) 10035-10043.
35. P. Babar, K. Patil, D.M. Lee, V. Karade, K. Gour, S. Pawar, J.H. Kim, Cost-effective and efficient water and urea oxidation catalysis using nickel-iron oxyhydroxide nanosheets synthesized by an ultrafast method. Journal of Colloid and Interface Science, 584 (2021) 760-769.
36. Y. Gong, H. Zhao, D. Ye, H. Duan, Y. Tang, T. He, L.A. Shah, J. Zhang, High efficiency UOR electrocatalyst based on crossed nanosheet structured FeCo-LDH for hydrogen production. Applied Catalysis A: General, 643 (2022) 118745.

37. Q. Zhang, M. Sun, M. Yao, J. Zhu, S. Yang, L. Chen, B. Sun, J. Zhang, W. Hu, P. Zhao, Interfacial engineering of an FeOOH@ Co_3O_4 heterojunction for efficient overall water splitting and electrocatalytic urea oxidation. Journal of Colloid and Interface Science, 623 (2022) 617-626.
38. Y. Wang, Z. Chen, H. Wu, F. Xiao, E. Cao, S. Du, Y. Wu, Z. Ren, Self-assembly-induced mosslike Fe_2O_3 and FeP on electro-oxidized carbon paper for low-voltage-driven hydrogen production plus hydrazine degradation. ACS Sustainable Chemistry & Engineering, 6 (2018) 15727-15736.
39. F. Yang, X. Chen, Z. Li, D. Wang, L. Liu, J. Ye, Ultrathin FeP nanosheets as an efficient catalyst for electrocatalytic water oxidation: Promoted intermediates adsorption by surface defects. ACS Applied Energy Materials, 3 (2020) 3577-3585.
40. X.-Y. Meng, M. Wang, Y. Zhang, Z. Li, X. Ding, W. Zhang, C. Li, Z. Li, Superimposed OER and UOR performances by the interaction of each component in an Fe–Mn electrocatalyst. Dalton Transactions, 51 (2022) 16605-16611.
41. T. Guo, X. Xu, X. Wang, J. Zhou, H. Wang, Z. Shi, M. Huang, Enabling the full exposure of Fe_2P@ NixP heterostructures in tree-branch-like nanoarrays for promoted urea electrolysis at high current densities. Chemical Engineering Journal, 417 (2021) 128067.
42. X. Xu, C. Zhang, J. Li, H. Liu, G. Su, Z. Shi, M. Huang, Redistributing interfacial charge density of $Ni_{12}P_5/Ni_3P$ via Fe doping for ultrafast urea oxidation catalysis at large current densities. Chemical Engineering Journal, 452 (2023) 139362.
43. H. Wang, S. Tao, Fabrication of a porous NiFeP/Ni electrode for highly efficient hydrazine oxidation boosted H_2 evolution. Nanoscale Advances, 3 (2021) 2280-2286.
44. J.S. Samdani, J. Sanetuntikul, S. Shanmugam, Self-supported iron-doped nickel sulfide as efficient catalyst for electrochemical urea and hydrazine oxidation reactions. International Journal of Hydrogen Energy, 47 (2022) 27347-27357.
45. D. Li, W. Wan, Z. Wang, H. Wu, S. Wu, T. Jiang, G. Cai, C. Jiang, F. Ren, Self-derivation and surface reconstruction of Fe-doped Ni_3S_2 electrode realizing high-efficient and stable overall water and urea electrolysis. Advanced Energy Materials, 12 (2022) 2201913.
46. W. Liu, E. Hu, H. Jiang, Y. Xiang, Z. Weng, M. Li, Q. Fan, X. Yu, E.I. Altman, H. Wang, A highly active and stable hydrogen evolution catalyst based on pyrite-structured cobalt phosphosulfide. Nature Communications, 7 (2016) 10771.
47. E. Budiyanto, S. Salamon, Y. Wang, H. Wende, H. Tüysüz, Phase segregation in cobalt iron oxide nanowires toward enhanced oxygen evolution reaction activity. JACS Au, 2 (2022) 697-710.

48a. C. Liu, D. Jia, Q. Hao, X. Zheng, Y. Li, C. Tang, H. Liu, J. Zhang, X. Zheng, P-doped iron–nickel sulfide nanosheet arrays for highly efficient overall water splitting. ACS Applied Materials & Interfaces, 11 (2019) 27667-27676.

48b. S.A. Khalate, S.A. Kadam, Y.-R. Ma, S.B. Kulkarni, V.G. Parale, U.M. Patil, Binder free cobalt iron phosphate thin films as efficient electrocatalysts for overall water splitting. Journal of Colloid and Interface Science, 613 (2022) 720-732.

49. B. Zhong, B. Cheng, Y. Zhu, R. Ding, P. Kuang, J. Yu, Hierarchically porous nickel foam supported Fe-Ni_3S_2 electrode for high-current–density alkaline water splitting. Journal of Colloid and Interface Science, 629, (2023) 846-853.
50. H. Yang, Y. Zhang, F. Hu, Q. Wang, Urchin-like CoP nanocrystals as hydrogen evolution reaction and oxygen reduction reaction dual-electrocatalyst with superior stability. Nano Letters, 15 (2015) 7616-7620.

51. Y. Liu, X. Liang, L. Gu, Y. Zhang, G.D. Li, X. Zou, J.S. Chen, Corrosion engineering towards efficient oxygen evolution electrodes with stable catalytic activity for over 6000 hours. Nature Communications, 9 (2018) 2609.

52a. E. Isarain-Chavez, M.D. Baro, C. Alcantara, S. Pane, J. Sort, E. Pellicer, Micelle-assisted electrodeposition of mesoporous Fe–Pt smooth thin films and their electrocatalytic activity towards the hydrogen evolution reaction. ChemSusChem, 11 (2018) 367-375.

52b. L. Han, S. Dong, E. Wang, Transition-metal (Co, Ni, and Fe)-based electrocatalysts for the water oxidation reaction. Advanced Materials, 28 (2016) 9266-9291.

53. H. Yang, M. Driess, P.W. Menezes, Self-supported electrocatalysts for practical water electrolysis. Advanced Energy Materials, 11 (2021) 2102074.

54. C. Chen, P. Zhang, M. Wang, D. Zheng, J. Chen, F. Li, X. Wu, K. Fan, L. Sun, Boosting electrocatalytic water oxidation by creating defects and lattice-oxygen active sites on Ni-Fe nanosheets. ChemSusChem, 13 (2020) 5067-5072.

55. J.M.V. Nsanzimana, Y. Peng, Y.Y. Xu, L. Thia, C. Wang, B.Y. Xia, X. Wang, An efficient and earth-abundant oxygen-evolving electrocatalyst based on amorphous metal borides. Advanced Energy Materials, 8 (2018) 1701475.

56. H. Su, W. Zhou, W. Zhou, Y. Li, L. Zheng, H. Zhang, M. Liu, X. Zhang, X. Sun, Y. Xu, In-situ spectroscopic observation of dynamic-coupling oxygen on atomically dispersed iridium electrocatalyst for acidic water oxidation. Nature Communications, 12 (2021) 6118.

CHAPTER

2

Advances in $Mg_3(Sb, Bi)_2$-based Zintl Phases for Thermoelectrics Energy Conversion

Nagendra S. Chauhan[1,2*] and Qiang Zhang[3]

[1] International Center for Materials Nanoarchitectonics (WPI-MANA), National Institute for Materials Science (NIMS), Namiki 1-1, Tsukuba, 305-0047, Japan

[2] Department of Applied Physics, Graduate School of Engineering, Tohoku University, Sendai, Miyagi, 980-8579, Japan

[3] Key Laboratory of Interface Science and Engineering in Advanced Materials, Ministry of Education, Taiyuan University of Technology, Taiyuan 030024, China

1. Introduction

Wasted heat accounts for over 60% of primary energy consumption globally, presenting an enormous and emerging prospective for its recovery through thermoelectric based power generation [1,2]. As a heritage technology for space applications, Radioisotope Thermoelectric Generators (RTGs) were widely employed in numerous space missions, including those involving deep space probes like Voyager and Cassini to ensure sustained power supplies in challenging extraterrestrial conditions [3,4]. Although terrestrial applications of thermoelectrics (TE) currently occupy only niche spaces like miniaturized cooling through peltier coolers, recent advancements and emerging landscape of TE technology, shown in Figure 1, is indicative of its forthcoming contribution to large-scale sustainable power generation [1,2]. Remarkably, TE based energy conversion being a carbon neutral process can contribute significantly to both primary energy generation using heat sources like solar or geothermal energy and energy conservation through the recovery of waste heat from automotive engines or nuclear power plants.

*Corresponding author: nagendra599@gmail.com; chauhan.nagendrasingh@nims.go.jp

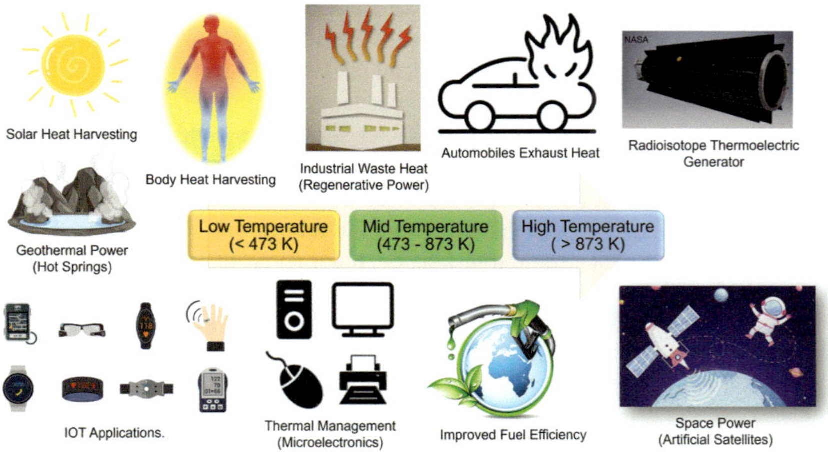

Figure 1. Emerging landscape of thermoelectrics.

The Seebeck effect, named after physicist Thomas Johann Seebeck, is a key phenomenon in TE conversion, which occurs when a temperature gradient (ΔT) is applied across a material, causing charge carriers to diffuse, generating an electric potential. Seebeck coefficient, which varies depending on the charge carrier transport, indicates whether the material is *p*-type or *n*-type, affecting the direction of current flow. Remarkably, solid-state TE devices enables direct interconversion of heat to electrical power via integrated *n*- and *p*-type semiconducting materials, thereby providing a motionless and reliable alternative to conventional heat engines. As building blocks, TE materials with both higher performance metrics and cost-effectiveness are essential to fully capitalize on TE prospects for waste heat recovery. The TE figure of merit i.e. zT, estimated as $zT = \left(\frac{S^2 \sigma}{\kappa}\right) T$, from the temperature dependent Seebeck coefficient (S), electrical conductivity (σ), and thermal conductivity (κ) at absolute temperature (T), is conventionally used to assess the maximum energy conversion efficiency of a TE device [5]. In a scenario where zT is infinitely high, the TE conversion cycle achieves reversibility (i.e. Carnot efficiency) for a device operating with hot side (T_h) and cold side (T_c) as evaluated by equation [5].

$$\eta = \frac{\Delta T}{T_h} \frac{\sqrt{1 + zT} - 1}{\sqrt{1 + zT} + \frac{T_c}{T_h}}$$

Although energy conversion efficiencies of current TE generator remain under 10% as indicated in Figure 2(a), primarily due to the prevalent average zT values ≈ 1 in state-of-the-art TE materials and their devices. Ongoing efforts are therefore aimed towards improving the zT value in a wider temperature, particularly at lower temperatures where low-grade heat sources are available in abundance. While high zT values are preferable in both p-type or n-type materials, many high-performance TE materials that are typically degenerate semiconductors are only available at higher temperatures [5]. The optimized TE materials typically exhibit a large S characteristic of semiconductors with narrow band gaps, in conjunction with ultra-low lattice thermal conductivity (κ_L) and moderately high σ. However, optimizing zT is complex due to the interrelation of transport properties as understood within the framework of solid-state physics. A heavy reliance on electronic transport i.e. holes (p-type) and electrons (n-type) becomes pivotal and important consideration for zT optimisation as illustrated in Figure 2(b) via. carrier concentration (n) dependent plot for TE parameters [5]. Fundamentally, even a superior material can exhibit a lower zT if it is significantly over- or under-doped.

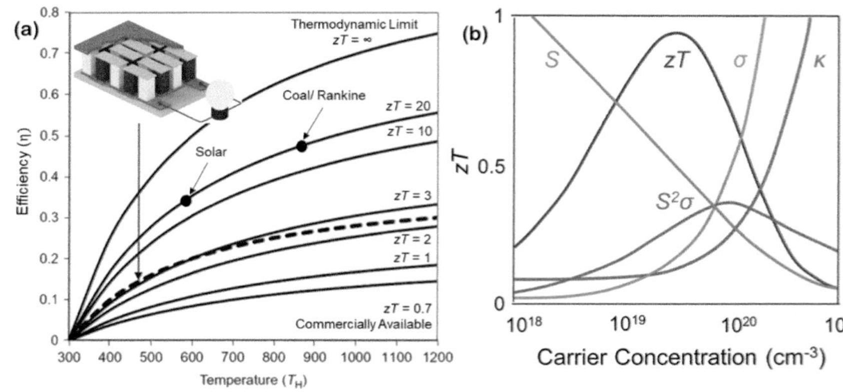

Figure 2. Thermoelectric efficiency (η) against hot-side temperature (T_H) for various zT values, with the cold side fixed at 300 K [6]. Conventional thermoelectric systems often exhibit efficiencies below 15% within the temperature range of 300–1000 K [5].

Good TE materials are often heavily doped semiconductors with a small band gap, which has an optimal n at around 10^{19}-10^{21} cm^{-3}. Alternatively, reduction of κ_L achieved by increasing phonon scattering primarily through alloying, defects and nanostructuring approaches, are extensively explored to improve the TE performance. The κ_L can be described as the product of heat capacity, phonon velocity and phonon mean free path. The reduction of phonon velocity is an effective way to decrease κ_L, while Umklapp scattering is often the dominant scattering mechanism at temperatures above the Debye temperature and materials with high Grüneisen parameters have intrinsically low κ_L. The reduction of κ_L is a key focus in materials which exhibits excellent power factors such as half-Heuslers and silicides. Steady improvements in zT on the material level alongside proliferating applications involving waste heat harvesting and powering wearable electronics has provided an expanding landscape for deployment of TE products in real world applications. Maximizing zT will enable capturing and productively utilizing a greater fraction of wasted thermal energy across automobiles, heavy manufacturing facilities, power plants and other systems that eject abundant excess heat. Meanwhile, assembling a toolkit of materials compatible with the temperatures, thermal fluxes, mechanical stresses, and geometrical constraints that are endemic to diverse working environments will facilitate their broader integration.

2. Overview of Zintl Phases in Thermoelectric Applications

Zintl phases represent a subset of intermetallic compounds formed between elements with significant electronegativity differences, most commonly alkali/alkaline earth metal and p-block element(s) as enlisted in Table 1 [7]. They are distinct from alloys, as the elements have clear oxidation states and undergo complete electron transfer to attain stable electronic configurations for both cations and anions, adhering to valence rules. Zintl phases demonstrate bond lengths that fall between metallic and ionic counterparts, exhibiting higher melting points compared to their elemental counterparts, and require a high heat of formation. In terms of mechanical properties, they tend to be more brittle than metals and display salt-like characteristics. These distinctive properties set them apart from other types of intermetallic compounds. The anionic networks in Zintl phases are characterized by covalent bonding, while the cationic component exhibits ionic characteristics. The covalent-anionic structure can adopt various topologies, including zero-dimensional isolated ions, dimers and polyatomic anions. Additionally, Zintl phases can form one-, two- and three-dimensional chains, planes and nets.

The Zintl-Klemm concept, established by Eduard Zintl and William Klemm [8], is a fundamental principle (with some exceptions) in understanding the structure, chemical bonding and properties of these compounds that have a pronounced ionic bonding character. Conceptually, ionic cations are mostly viewed as electron donors, which contribute their electrons to the covalently bonded anionic substructures. Being made up of a polyanionic substructure and group 1 or 2 counter cationic ions, structure in a Zintl compound can be understood by a formal electron transfer from the electropositive element to the more electronegative element in their composition. This concept is essential for rationalizing the electronic transport and provides qualitative insights into the properties of Zintl compounds. However, it is important to note that the Zintl formalism and electron counting are not universal and can only offer a qualitative understanding. Nevertheless, Zintl compounds and their formalism provides a pathway for the design of new compounds with diverse technological applications, allowing for the rational design and optimization of materials with properties such as thermoelectricity, magneto-electronics, photovoltaics and energy storage capabilities. Recently, these phases have also gained recognition for their roles as topological insulators and Dirac semimetals, highlighting their expanding significance in the realm of quantum properties.

Table 1. Zintl Phases for thermoelectric applications [7]

Generic Chemical Formula	Structure Type	Space group	Examples
$A_{11}B_6X_{12}$	$Sr_{11}Cd_6Sb_{12}$ [9]	C2/m	$Eu_{11}Cd_6Sb_{12}$ [10]; $Eu_{11}Zn_6Sb_{12}$ [10]
$A_9B_{4+x}X_9$	$Ca_9Mn_{4+x}Bi_9$	Pbam	$Yb_9Mn_{4.2}Sb_9$ [11]; $Yb_9Zn_{4.2}Sb_9$ [12]
AB_2X_2	$CaAl_2Si_2$	P-3m1	$YbZn_2Sb_2$ [13,14]; $YbMn_2Sb_2$ [15,16]; $YbCd_2Sb_2$ [14]; $EuZn_2Sb_2$ [17]
A_2BX_2	Yb_2CdSb_2	$Cmc2_1$	Ca_2CdP_2; Ba_2CdAs_2; Sr_2CdAs_2
$A_{14}BX_{11}$	$Ca_{14}AlSb_{11}$	$I4_1/acd$	$Yb_{14}MnSb_{11}$ [18,19]

3. Significance of Mg_3Sb_2 and Its Analogues in Thermoelectrics

Promising advancements in Zintl phases as TE materials has been demonstrated in recent years, primarily attributed to their inherently low lattice κ_L and highly tunable electronic properties [20–23]. From the phonon glass electron crystal (PGEC) concept viewpoint, Zintl phases are ideal candidates for efficient TE materials, because they are typically small bandgap semiconductors with complex structures [24–27]. Furthermore, such phases allow fine adjustment of dopant concentration without disrupting the carrier's mobility, which is essential for optimizing TE material efficiency. The structural complexity of many Zintl compounds leads to near-zero optical phonon mode velocity, while their covalently-bonded polyanionic structures contribute to band gap formation and in some cases, reasonable electronic mobility [28]. Thus, many Zintl compounds with high zT have been reported in the past few decades such as $Yb_{14}MnSb_{11}$ [18,19], $Yb(Zn, Cd)_2Sb_2$ [13], $Eu_5In_2Sb_6$ [29], $(Eu_{0.5}Yb_{0.5})_{1-x}Ca_xMg_2Bi_2$ [30], and $Yb_9Mn_{4.2}Sb_9$ [11] by manipulating the crystalline order, tunable electronic properties via aliovalent substitutions and intrinsically low thermal conductivities owing to their structural and chemical complexity. However, most of these compositions constitutes of expensive rare earth elements (Yb, Eu) and/or toxic elements (Cd), limiting their industrial power generation applications.

The choice of elemental constituents in high performance Zintl phases had raised concerns about their applicability and scalability due to material toxicity and high cost, necessitating a balance between performance, cost, and environmental impact. Simultaneously, there is a growing interest in materials made from abundant elements, processed at lower temperatures, and with the potential for large-scale production. In this context, among reported Zintl compounds in TE studies, $Mg_3(Sb, Bi)_2$ which constitutes of cheap, environmentally friendly and earth abundant elements, has garnered increasing attention and will be discussed primarily in this chapter. The realization of superior *n*-type TE performance in Te-doped *n*-type $Mg_3(Sb, Bi)_2$ as shown in Figure 3, with an extraordinarily high $zT > 1.6$ @ 773 K, has established their remarkable prospects for both power generation and cooling applications [20–23]. The TE properties of Mg_3Sb_2 have been extensively characterized in recent years, revealing peak power factor around 20 $mWcm^{-1}K^{-2}$ and inherently low $\kappa_L \sim$ 1-2 $Wm^{-1}K^{-1}$, which is comparable to the state-of-the-art Bi_2Te_3 [20,32]. The optimization of $Mg_3(Sb, Bi)_2$ TE performance through various methods such as controlling the Mg stoichiometry, [33,34] carrier scattering mechanism [21,35], doping [36,37], defect engineering [22,38] and increasing grain size [39–41] has been successfully applied to mitigate undesirable effects and improve TE performance, thereby achieving a state-of-the-art average *zT* value up to 1.1 in the range of 300-500 K [20–23]. These advancements are significant steps towards the realization of terrestrial applications of TE materials, particularly in wasted heat recovery technology.

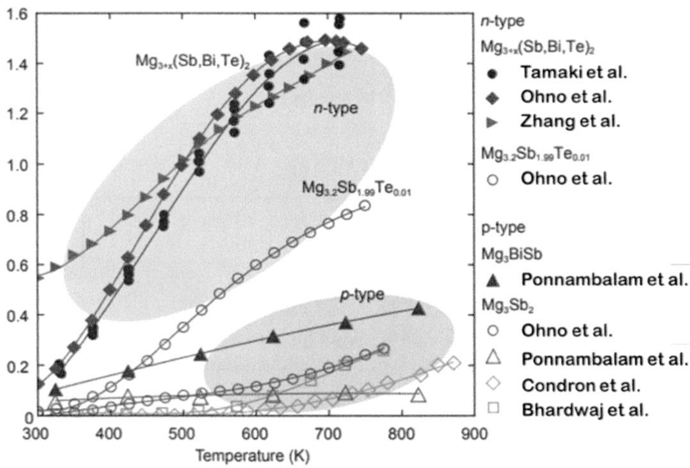

Figure 3. Comparison of the thermoelectric figure of merit (*zT*) between *n*- and *p*-type Mg_3Sb_2-based compounds that highlights the superior properties of *n*-type materials. (Reproduced with permission from Ohno et al. [31] Copyright 2017 Published by Elsevier Inc.)

4. Crystal Structure and Bonding

$Mg_3(Sb, Bi)_2$ have garnered significant interest in recent years for both low- and mid-temperature range TE applications [20–23]. As a Zintl phase material first reported [42] in 1933, Mg_3Sb_2 and its structural analogue Mg_3Bi_2 are noteworthy members of the 1-2-2 Zintl family, exhibiting $CaAl_2Si_2$-type structure [43]. They crystallize in a trigonal anti α-La_2O_3 structure with space group Pm1, wherein Mg atoms occupies two sites, denoted as Mg(1) and Mg(2) respectively as shown schematically in Figure 4 [20]. Mg(1) site forming a Mg^{2+} cation layer is octahedrally coordinated to Sb/Bi atoms, while Mg(2) site is tetrahedrally coordinated with Sb/Bi, forming the $[(MgSb)_2]^{2-}$ anionic sub-structure [7,20,25]. In high-temperature cubic phase (space group Ia-3) above 1193 K of Mg_3Sb_2 (which is not yet observed for Mg_3Bi_2), Mg occupies tetrahedral sites only. The evaluated instability of the octahedral site Mg_3Sb_2 is linked to the absence of tetrahedral sites in the cubic high-temperature phase [44].

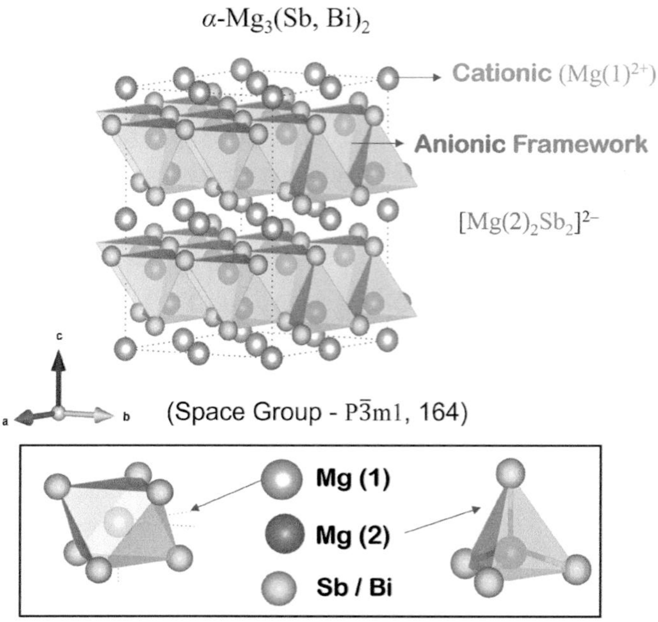

Figure 4. Depiction of the crystal structure of Mg_3Sb_2 in perspective view, to represent a unit cell with exhibiting $CaAl_2Si_2$-type structure. Also, Mg(1) octahedral and Mg(2) tetrahedral coordination is shown below.

Based on electron density topology analysis, it was proposed that the interactions between Mg and Sb atoms in Mg_3Sb_2 are predominantly ionic, wherein the calculated atomic charges, employing Bader partitioning, closely align with the formal charges of Mg^{2+} and Sb^{3-} (+1.5 on Mg and –2.2 on Sb), inferring strong ionicity in the Mg_3Sb_2 compound [45]. However, this perspective was countered by highlighting the significance of covalent interactions between Mg and Sb, an example of Mg-pnictides through Electron Localization Function (ELF) analysis [46], revealing basins between Mg and Sb atoms, that were indicative of covalent bonding. Additionally, their calculations, incorporating integrated Crystal Orbital Hamilton Population (COHP), yielded a value of around −6 eV over Mg and all its nearest neighbour Sb atoms, further supporting a covalent interpretation. The band resolved projected COHP for the interaction between Mg octahedra (Mg1) and Mg tetrahedra (Mg2) in Mg_3Sb_2 indicate anti-bonding states (positive values) between Mg atoms at the Valence Band Maximum (VBM), while bonding character (negative values) with interactions between 3*s* levels of Mg(1) and Mg(2) was evaluated at the Conduction Band Minimum (CBM) [47]. From a structural and application point of view, diverse structural chemistry, unique bonding states and covalent interaction in Mg_3Sb_2 provides an additional freedom for band manipulations compared to other Zintl compounds with purely ionic interactions.

Moreover, in accordance with Pauling's sphere-packing model, the smallest stable cation-to-anion radius ratio for octahedral coordination is given as r_{cation}: r_{anion} = 0.414 [49]. In the case of AMg_2X_2 compounds with $CaAl_2Si_2$ type structure, this criterion is met for Ca, Sr, Eu, and Yb as shown in Figure 5 which illustrates the estimated r_{cation}: r_{anion} [48]. However, the size of Mg, is smaller than any other cation that can occupy the octahedral site (e.g., Ca, Yb, Sr, Eu, and Ba) resulting in a distorted octahedral environment [48]. Interestingly, for A = Mg the r_{cation} : r_{anion} values are below the stability limit, while compounds with larger cations are anticipated to be stable in a 6-fold coordinated environment. Thus, the smaller size of the Mg atom at the octahedral site, violates the stability limit, weakens interlayer bonding, leading to soft shear modes and a large Grüneisen parameter. Interestingly, Mg_3Sb_2 exhibits an inherent structural instability in its seemingly simple crystal structure, suggesting a remarkably high phonon scattering rates and consequently lower κ_L, as discussed in subsequent sections.

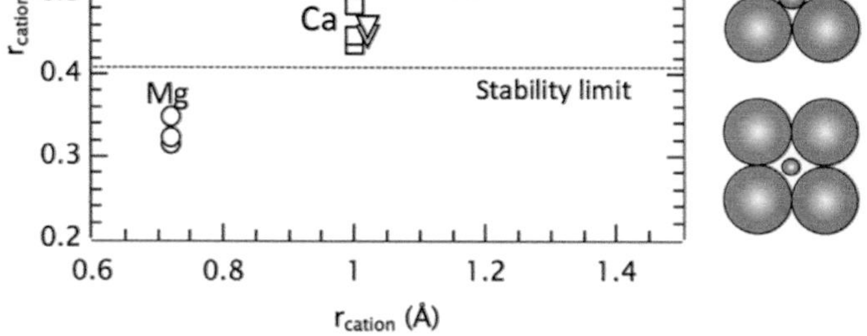

Figure 5. Cation : Anion ratio stability limit for AMg_2X_2 compounds with $CaAl_2Si_2$ type structure. (Reproduced with permission from Peng et al. [48] Copyright 2018, John Wiley & Sons, Inc.)

5. Synthesis Complexity and Phase Boundary Mapping

The synthesis of Zintl phases to achieve the desired compositions and crystal structures is often complex and requires controlled conditions and stoichiometry during synthesis. Interestingly, semiconductors with more than two elements can exhibit different properties depending on thermodynamic growth conditions. For instance, in a binary semiconductor AB, two distinct semiconductors may form at a given temperature—one with A-excess and the other with B-excess. Thus, different synthesis conditions can lead to variations in charge carrier concentration and may result in *n*-type or *p*-type semiconductors. In the context of $Mg_3(Sb, Bi)_2$ research, phase diagrams of Mg_3Sb_2 [50] and Mg_3Bi_2 [44] shown in Figure 6, with existing phases tabulated in Tables 2 and 3, respectively indicates the importance of phase boundary mapping in realising the desired phase, as overcoming/tuning the intrinsic *p*-type behaviour and phase stability in Mg_3 (Sb, Bi)$_2$ have remained a formidable challenge during synthesis.

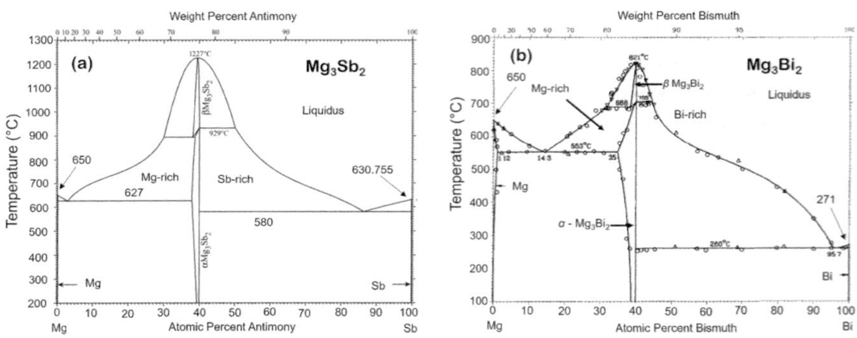

Figure 6. Phase diagram for (a) Mg_3Sb_2 [50] and (b) Mg_3Bi_2 [44].

Excess Mg is essential for achieving *n*-type characteristics in the Mg_3(Sb, Bi)$_2$, implying the potential involvement of interstitial Mg. However, defect chemistry of Mg_3Sb_2 through theoretical and experimental approaches elucidate the existence of two distinct thermodynamic states (Mg-excess and Sb-excess), wherein only one i.e., Mg-excess can exhibit *n*-type behaviour [31]. In a typical 'line compound,' multiple thermodynamic states significantly impact electronic properties by suppressing unwanted defects differently. To create the desired *n*-type Mg_3Sb_2-based compound, it is critical to favour the Mg-excess thermodynamic state, thereby introducing a synthesis strategy termed 'phase boundary mapping,' which involves adding excess constituents until the impurity phase is identified. Contrary to conventional perspectives favouring thermodynamic phase purity, it allows for the discovery of various versions of a compound, even those considered 'line compounds' with no measurable compositional variation [51,52]. Through,

Table 2. Mg-Sb crystal structure data [50]

Phase Composition	at.% Sb	Pearson symbol	Space group	Prototype
(Mg)	0	hP2	$P6_3/mmc$	Mg
β-Mg_3Sb_2	38.1-40	cI80	Ia	Mn_2O_3
α-Mg_3Sb_2	37.9-40	hP5	P-3m1	La_2O_3
(Sb)	100	hR2	R-3m	α-As

Table 3. Mg-Bi crystal structure data [44]

Phase Composition	at.% Sb	Pearson symbol	Space group	Prototype
(Mg)	0	hP2	$P6_3/mmc$	Mg
β-Mg_3Bi_2	40	-	-	-
α- Mg_3Bi_2	40	hP5	P-3m1	La_2O_3
(Bi)	100	hR2	R-3m	α-As

phase boundary mapping exploration of different thermodynamic states of the compounds is possible by crossing phase boundaries, without requiring precise information about the exact location or specific composition of phase boundaries.

Remarkably, ball milling methods have demonstrated greater efficacy compared to melting techniques like high-temperature furnace melting and arc melting. High reactivity of Mg with quartz ampoules and the elevated Mg vapor pressure during synthesis, often results in formation of intrinsic defects such as Mg vacancies [45]. The persistent formation of Mg vacancies during high temperature synthesis, results in the generation of excess holes (h^+) as V_{Mg}^{2-} ($Mg_{Mg} \Rightarrow V_{Mg}^{2-} + 2\,h^+$) [33]. Thus, for the synthesis of $Mg_3(Sb, Bi)_2$ based compounds, attaining *n*-type conduction was a synthesis obstacle, as extensively reported in preliminary studies.[53–56] This challenge was overcome by introducing excess Mg coupled with Te doping, thereby attaining desired *n*-type electrical transport in the synthesized materials. Following which, *n*-type Mg_3Sb_2-based materials have been synthesized primarily by ball-milling [20,21,35] methods as indicated in Table 4, besides melting methods [45], reactive sintering [57–59] and single crystals growth [60–62].

Mechanical alloying via. ball milling ensures homogeneity in reacted powders, with a rapid sintering (hot press/spark plasma sintering) applied post-milling for completing solid-state reaction and powder densification. Thus, for addressing Mg_3Sb_2 reactivity and high Mg vapor pressure, high energy ball milling offers accurate composition control, shorter reaction time and homogeneous temperature distribution. The processing duration ranges from 1-10 hour, while sintering temperatures from 873–1073 K in diverse studies as enlisted in Table 4. The powder obtained through ball milling exhibits flammability upon rapid exposure to air. Therefore, it is imperative to handle powder samples under an

Table 4. Room-temperature transport properties of Mg_3Sb_2 synthesized by different methods

Synthesis*	*R*	*n*	*S*	μ	κ_L	References
	Ω.cm	$10^{19}cm^{-3}$	μVK^{-1}	$cm^2V^{-1}s^{-1}$	$Wm^{-1}K^{-1}$	
Arc melting	0.002	2	-200	91.8	0.7-0.9	Zhang et al. [45]
Melting in Mo crucible	2.520	0.02	400	11.3	-	Kim et al. [60]
Melting in Ta tubes	3.3	-	363	-	2.08	Ahmadpour et al. [63]
Melting in quartz tubes	0.150	0.8	50	10	4.6	Ponnambalam et al. [56]
Ball milling	0.0028	12.1	100	39.3	-	Ren et al. [64]
HP (1213 – 1273 K) @ 24 MPa	0.08	-	250	-	2.0	Kajikawa et al. [55]
SSR (1073 K – 7 Days) + HP (873 K @ 138 MPa)	50	-	25	-	4.2	Condron et al. [65]
BM (20 hrs)+ HP (1023 K)	-	0.00026	1000	23.2	1.4	Shuai et al. [66]
BM – SPS (1123 K @ 75 MPa)	0.105	0.175	320	28.3	1.4	Song et al. [67]
SPS (973 K @ 40 MPa)	0.088	0.087	-550	82	1.8	Tani et al. [57,58]

* BM – Ball Milling; SPS – Spark Plasma Sintering; HP – Hot Pressing

inert atmosphere to ensure safety and prevent potential combustion hazards. The melting temperature is higher in Mg_3Sb_2 (above 1473 K) [50] and Mg_3Bi_2 (1093 K) [44] as evaluated from their respective phase diagrams shown in Figure 6. At 773 K, the vapor pressures of elemental Mg, Sb, and Bi are 7,390 Pa, 139 Pa and 9.64 Pa, respectively. The considerable difference in vapor pressure among these constituents' elements suggest a significant Mg loss during sintering, particularly at high temperatures, as observed extensively in various studies. Through a meticulous comparison of electrical conductivity in various reports, it becomes evident that synthesis and processibility do play a dominant role in tuning the TE transport, as observed in these Zintl compounds.

Supersaturated Mg vacancies exhibit diffusion within the crystal lattice during the sintering procedure, ultimately dissipating as they permeate into the grain boundaries or surfaces of the specimens. Besides, intrinsic Mg vacancies (which represent thermodynamic equilibrium defect), the ball-milling method also introduces a substantial quantity of supersaturated Mg vacancies into *n*-type Mg_3Sb_2-based materials. However, successful reduction of Mg-vacancy concentration and the achievement of *n*-type conduction, can be attained through the incorporation of additional Mg into the matrix. Moreover, elevated sintering temperatures and prolonged holding time, explicates the heightened carrier mobility, as observed in studies for systematic optimization of preparation conditions. Furthermore, the control of supersaturated Mg vacancies through extended high-temperature annealing at elevated temperatures, although seemingly feasible, is likely to induce a reaction between Mg_3Sb_2 and the quartz tube, thus precautions must be implemented to prevent such reactions. Thus, maximizing control over the crystal structure, grain size, orientation, and interfaces, remains the way forward and can be achieved through innovative synthetic approaches.

6. Electronic Band Structure

The electronic transport properties of Mg_3Sb_2-based materials have garnered increasing attention, especially in recent years, with a focus on optimizing both *p*-type and *n*-type electrical properties through band engineering and orbital degeneracy tuning. The exceptional electronic conduction properties observed in *n*-type Mg_3Sb_2, were characterized by a substantial N_v/m^* value and isotropic features, where N_v represents the number of valleys and m^* denotes the band's effective mass. Using first-principles methods and Boltzmann transport theory, the evaluated electronic structure of Mg_3Sb_2 reveals low-lying conduction band structure with multi valley character and light effective mass of *n*-Mg_3Sb_2 ($N_v = 6$, $m^* = 0.19\ m_e$), leading simultaneously to a high S and a good carrier mobility, as indicated in Figure 7(a) [20]. The pioneering studies suggested Mg_3Sb_2 as a semiconductor (with a band gap of approximately 0.5 eV), while Mg_3Bi_2 to be a semimetal as shown in Figure 7(b) [53–56,68,69]. The six-fold degenerate conduction band minimum at low symmetry k points shown in Figure 7(c), significantly contribute to its superior *n*-type TE performance compared to *p*-type materials. The evaluated N_v/m^* for *n*-type Mg_3Sb_2 is also found to be comparable to the benchmark TE materials like Bi_2Te_3 and PbTe as shown in Table 5, showcasing its potential. Encompassing both higher energy Mg-*p* orbitals and lower energy Sb-*s* orbitals, besides the Mg-*s*–Sb-*p* orbital interactions, are essential for comprehending the configuration of the lowest conduction band and its alteration upon alloying. The calculated bulk modulus (B), shear modulus (G), Young's modulus (E) and Poisson ratio (υ) evaluates that Mg_3Sb_2 and Mg_3Bi_2 phases are both ductile [70].

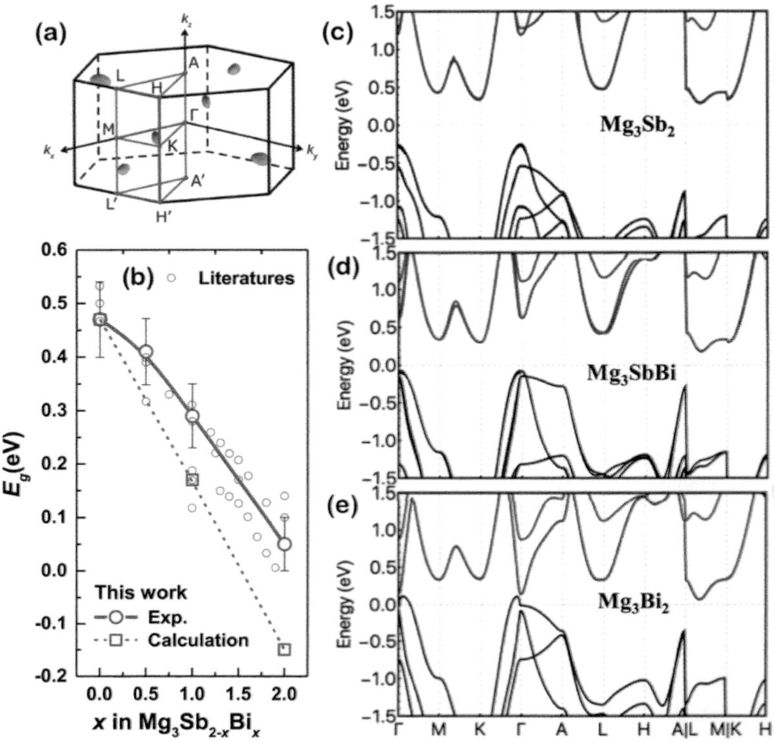

Figure 7. (a) Fermi surface of the conduction band exhibits six equivalent carrier pockets, situated near the midpoint between M and L'. (Reproduced with permission from Tamaki et al. [20] Copyright 2016, John Wiley & Sons, Inc.) (b) Composition-dependent band gap (E_g) for $Mg_3Sb_{2-x}Bi_x$ alloys is shown, along with a comparison to literature results, and calculated band structures are presented for (c) Mg_3Sb_2, (d) Mg_3SbBi, and (e) Mg_3Bi_2. (Reproduced with permission from Shi et al. [69] Copyright 2021, Elsevier Inc.)

Table 5. Descriptor N_v/m^* values of Bi_2Te_3, PbTe, and Mg_3Sb_2 obtained by generalized gradient approximation. For Bi_2Te_3, a spin–orbit interaction is additionally taken into consideration [20]

Material	**Carrier type**	**Valley number (N_v)**	**Band effective mass (m^*)**	**Descriptor (N_v/m^*)**
Mg_3Sb_2	*n*	6	0.19	31.1
	p	1	0.17	5.9
Bi_2Te_3	*n*	6	0.17	35.2
	p	6	0.17	35.6
PbTe	*n*	4	0.19	21.2
	p	4	0.15	26.1

Lately, researchers have been examining the dopability through defect formation by using DFT calculating in search of potential materials with high TE efficiency [22,38]. The defect formation energy of a defect in a charge state can be computed employing DFT, wherein the smaller formation energies indicate greater favourability. If the defect formation energy becomes negative, the defect spontaneously forms. In Mg_3Sb_2-based materials, Mg vacancies also known as "killer defects" have the lowest defect formation energy around the Fermi level, limiting the increase in carrier concentration and causing *p*-type conduction as indicated in Figure 8. To counteract these killer defects, the appropriate choice of thermodynamic state in the phase diagram is critical and can be understood employing phase boundary mapping the concept. The materials design principle of *p*-Mg_3Sb_2, where the dominant anisotropic p_z orbital at Sb is influenced by the crystalline field effect [45], can also be understood and favorably tuned by controlling the defect formation around the Fermi level.

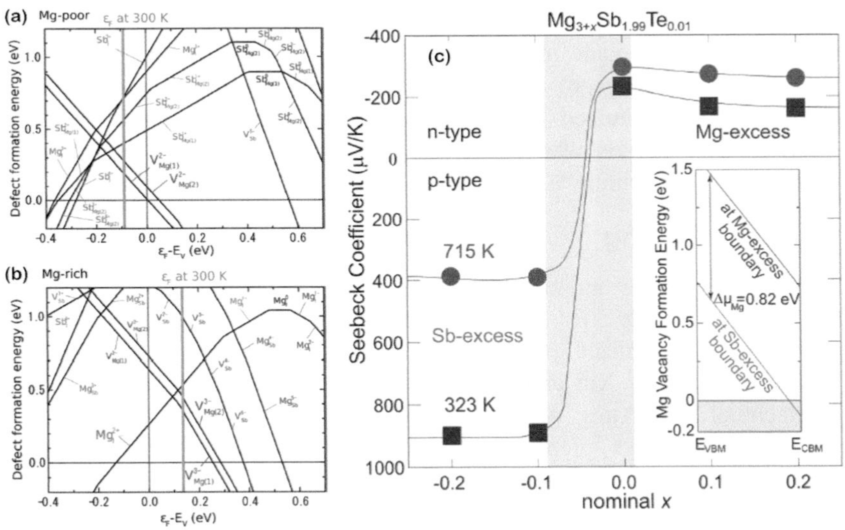

Figure 8. (a) Calculated Eform in the Mg-poor condition, representing the threshold for Sb segregation. (b) E form in the Mg-rich condition, corresponding to the solubility limit for excess Mg. The calculated E_F values align with the observation that stoichiometric Mg_3Sb_2 persists as a *p*-type material. (c) The successful incorporation of excess Mg proves effective in achieving *n*-type conduction upon doping. (Reproduced with permission from Ohno et al. [31] Copyright 2017 Published by Elsevier Inc.)

7. Carrier Concentration Optimization

Doping is a critical strategy in TE research extensively exploited for controlling/optimisation of carrier concentration. A common strategy to modulate carrier concentration involves introducing extrinsic dopants. By substituting at least one element in the compounds with aliovalent elements, electrons or holes can be introduced. However, intrinsic defects play a key role, acting as "killer defects" that hinder precise control of the Fermi level (i.e., *n*), thereby complicating the optimization strategies for *n*. Defects are pervasive in materials and naturally occur in various forms. Intrinsic defects, such as vacancies, interstitials, and antiseptics defects, as well as extrinsic defects, introduce a valence imbalance of electrons and holes, thereby modifying the transport of charge carriers. In this regard, Phase boundary mapping [31] provides a valuable insight into the dopability of Mg_3Sb_2, by understanding and controlling the defect formation through appropriate selection of thermodynamic state in the phase diagram. It also addresses and overcome, the typical consideration of "line compounds" that assume uniform properties for samples of the same composition. Among TE transport properties, resistivity shows a significant difference and inverse variation with sintering temperatures, with decreasing values for rising temperatures [22]. In this section, we present a succinct overview of successful dopants employed for *p*-type and *n*-type doping in Mg_3Sb_2, emphasizing the vital shortcomings in obtaining an adequate number of charge carriers. For *p*-type Mg_3Sb_2, the experimental challenge of obtaining an adequate number of charge carriers often leads to unoptimized TE properties, is supported by theoretical outcomes. Unlike *n*-type materials, where popular dopants like Te and Se were successfully incorporated for attaining a high electron concentration ($\gtrsim 10^{20}$ cm^{-3}).

7.1 *p*-type Doping

Undoped Mg_3Sb_2 is an intrinsic *p*-type semiconductor with a low carrier concentration of only ~10^{17} cm^{-3}. Despite exhibiting a very small κ_L, Mg_3Sb_2 has long been recognized as *p*-type material with modest TE performance. The pioneering study by Xin and Qin [71] explored the temperature dependence of electrical properties of the nano-crystal substitutional semiconductor $Mg_3(Bi_xSb_{1-x})_2$ alloys prepared by mechanical alloying and hot pressing, indicating transition from the semiconducting state to the metallic state with increasing Bi-content. Despite substantial theoretical and experimental efforts to explore efficient *p*-type dopants in Mg_3Sb_2, the high hole concentration ($\gtrsim 10^{20}$ cm^{-3}) for favorable *p*-type doping behaviour remains largely ineffective to date. While band structure engineering has emerged as a powerful tool for tailoring electronic and phononic properties, improving the poor carrier mobility in intrinsic *p*-type Mg_3Sb_2 still remains a formidable challenge. Different types of dopants/additives as illustrated in Table 6, have been investigated for increasing *n* through accepting dopant, such as Na [64, 66], graphene [72], Bi [73] Li [74, 75], Cd [75], Zn [64, 68], Al [76], and Ag [67, 77, 78]. Among the investigated Mg-site dopants, Na stands out as the most effective (yielding a high-power factor of 7.5×10^{-4} $Wm^{-1}K^{-2}$), while Pb doping on Sb-site has demonstrated an enhancement in $zT \sim 0.84$ at 773 K. These efforts to improve TE performance, have led to $zT = 0.6$-0.8 at mid-temperature around 750 K as enlisted in Table 6.

Table 6. Progress in doping *p*-type Mg_3Sb_2

Year	Compositions/Approach	zT_{max}	Processing*	References
2020	$Mg_3Sb_{2-x}Te_x$	1 (768 K)	SPS (973 K @ 40 MPa)	Tani et al. [58]
2020	$Mg_3Sb_2Y_x$	1.0 (773 K)	SPS (973 K @ 40 MPa)	Tani et al. [57]
2020	$Mg_{3.19-x}Li_{0.01}Cd_xSb_2$	0.68 (773 K)	BM (10 hrs) + HP (973 K @ 80 MPa)	Tang et al. [75]
2019	$Mg_{3+\delta}Sb_xBi_{2-x}$ (Bi-Doping)	-	BM (10 hrs) + SPS (948 K @ 50 MPa)	Shu et al. [37]
2018	$Mg_{3\text{-}x\text{-}0.0125}Zn_xNa_{0.0125}Sb_2$	0.8 (773 K)	BM (10 hrs)-HP (923 K @ 50 MPa)	Ren et al. [64]
2016	$Mg_{3-x}Zn_xSb_2$	0.37 (773 K)	BM + SPS (1073 K@ 50 MPa)	Bhardwaj et al. [68]
2015	$Mg_3Sb_{2-x}Bi_x$	0.94 (773 K)	BM + SPS (873 K@ 50 MPa)	Bhardwaj et al. [73]
2015	$Mg_3Sb_{1.8}Bi_{0.2}$/GNS	1.35 (773 K)	BM (15 hrs) + SPS (1073 K@ 50 MPa)	Bhardwaj et al. [72]
2015	$Mg_{3-x}Na_xSb_2$	0.6 (773 K)	BM (20 hrs)+ HP (1023 K)	Shuai et al. [66]
2015	$Eu_5In_{2-x}Zn_xSb_6$	0.4 (660 K)	SSR + BM + HP (823 K @ 80 Mpa)	Chanakian et al. [29]
2014	$Mg_3Sb_{2-x}Pb_x$	0.84 (773 K)	BM + SPS (1073 K@ 50 MPa)	Bhardwaj et al. [81]
2006	$Mg_3(Bi_xSb_{1-x})_2$	-	BM + HP (300 MPa @ 673 K)	Xin et al. [71]
2005	$Ca_xYb_{1\text{-}x}Zn_2Sb_2$	0.5 (673 K)	SSR + HP (723 K)	Gascoin et al. [13]
2003	Mg_3Bi_2		BM + HP (849 – 890 K) @ 24 MPa	Kajikawa et al. [55]

* BM – Ball Milling; SPS – Spark Plasma Sintering; HP – Hot Pressing; SSR – Solid State Reaction

The TE performance of *p*-type Mg_3Sb_2 is a critical consideration, as the zT values for this material generally remain below unity. The primary challenge lies in the poor electrical transport properties of *p*-type Mg_3Sb_2, significantly limiting its practical applications. Given that both *p*- and *n*-type legs are essential for power generation devices to maximize power output, enhancing the TE performance of *p*-type Mg_3Sb_2 becomes pivotal to improve its overall efficiency and applicability. It should be noted that there is a possibility of property anisotropy in *p*-type conduction due to the anisotropic crystal structure, as revealed by directional solidification experiments [77] and theoretical calculations [79,80]. Among TE transport properties, resistivity is particularly anisotropic, with a difference by a factor of approximately 4.5 between crystallographic directions along [001] and [101] in *p*-type Mg_3Sb_2 due to different carrier mobility and effective mass for different orientations. However, in *n*-type conduction, the difference in resistivities along [001] and [101] is less than 20% [80]. Furthermore, the potential for developing *p*-type TE materials, such as $CaMg_2Bi_{1.98}$, within the $CaAl_2Si_2$-type structure has been explored, further enhancing the TE performance of these materials. Furthermore, the $CaAl_2Si_2$-type structure has been investigated for its TE potential, with materials like $Ca_{1-x}Yb_xZn_2Sb_2$ showing promise for TE applications. These materials hold potential for low- and intermediate-temperature TE applications due to their favourable TE properties, making them promising candidates for energy conversion technologies.

7.2 *n*-type Doping

The *n*-type Mg_3Sb_2-based materials are well recognized as an outstanding TE materials owing to their remarkable *n*-type TE properties. To achieve high TE performance in *n*-type Mg_3Sb_2-based TE materials, various strategies such as high conduction band degeneracy [21,45], intrinsically high carrier mobility [23,36,82], grain boundary engineering [32,33], and phase boundary mapping [31] have been successfully demonstrated. A broad range of carrier density, tunability was reported for optimizing transport properties along with efficient *n*-type dopants i.e., blending of excess Mg and impurity doping with Te or Se, to overcome persistent Mg vacancies formed during synthesis. The Mg vacancies, also regarded as "killer defects", are characterized by low formation energy, that impedes the desired *n*-type conductivity characteristics, in the synthesized materials. Alongside various experimental endeavours as enumerated in Table 7, aimed at enhancing the low-temperature performance of *n*-type $Mg_{3+\delta}Sb_{2-x}Bi_xTe_y$, theoretical investigations have provided insights into accurate electronic structure, [14,15] chemical bonding, [16–18] and phonon transport properties [19,20]. A high electron concentration ($\gtrsim 10^{20}$ cm^{-3}) can be achieved by doping on the Mg sites with dopants such Mo, Nb, Ta, Y, Ge, Mn, Co, lanthanide dopants (Pr [83], Nd [83,84], Gd, Ho), which are generally more efficient than other and the anionic Sb-site dopant. On the Sb-site, Se and S have been reported as effective *n*-type dopants for Mg_3Sb_2, [14, 21] albeit less efficient than Te due to higher substitution defect formation energies. The charge density map of the conduction band highlights a 3*D* conduction network formed by cationic Mg^{2+} and anionic $[Mg_2Sb_2]^{2-}$, resulting in nearly isotropic effective mass values ($m_x = 0.20\ m_e$ and $m_z = 0.17\ m_e$) across six valleys favouring remarkable *n*-type Mg_3Sb_2-based TE materials [20,79]. The observed thermally activated behaviour with positive temperature dependency in the electrical conductivity/mobility resulted in low electrical performance around room temperature i.e. 300 K-500 K. Shifting the peak *zT* temperature or eliminating this activation behaviour could potentially optimize the system for use in the lower temperature range, where high-performance *n*-type materials are scarce.

Table 7. Progress in doping *n*-type Mg_3Sb_2

Year	Compositions / Approach	zT_{max}	Processing*	References
2023	$Mg_{3.1}X_{0.1}Sb_{1.5}Bi_{0.49}Te_{0.01}$ (X = Nb and Ta)	2.04 (798 K)	BM (8 hrs) + SPS (923K /1073 K @ 50 MPa)	Li et al. [85]
2023	$Mg_{3.2}Bi_{1.49}Sb_{0.5}Te_{0.01}Ni_x$		BM (12 hrs) + SPS (723K @ 20MPa)	Li et al. [86]
2023	$(Mg_2Sn_{0.85}Sb_{0.15})_{1-x}(Mg_3Sb_2)_x$	1.4 (773 K)	BM + SPS (873K @ 40 MPa)	Zhu et al. [87]
2022	$Mg_{3.2-x}Y_xSb_{1.5}Bi_{0.49}Se_{0.01}$	1.87 (773 K)	BM (5 hrs) + SPS (923 K @ 50 MPa)	Liang et al. [88]
2022	$Mg_{3.2}Sb_{1.5}Bi_{0.49}Se_{0.01}$	1.21 (723 K)	BM +SPS (923K @50 MPa)	Yu et al. [83]
	$Mg_{3.2-y}Pr_ySb_{1.5}Bi_{0.49}Se_{0.01}$	1.67 (723 K)		
	$Mg_{3.2-y}Nd_ySb_{1.5}Bi_{0.49}Se_{0.01}$	1.74 (723 K)		
2022	$Mg_{3.2}Bi_{0.5}Sb_{1.49-2x}Te_{0.01+x}Ge_x$ (Janus Nanoprecipitation)	1.6 (773 K)	BM (10 hrs)-HP (923 K @ 50 MPa)	Shu et al. [89]
2021	$Mg_{3+\delta}Nd_ySb_{2-x}Te_x$;	1.26 (725 K)	BM (15 min) + SPS (1073 K @ 60 Mpa)	Zhang et al. [90]
	$Mg_{3+\delta}Ho_ySb_{2-x}Te_x$	0.94 (725 K)		
2020	$Mg_{3+\delta}Tm_ySb_{2-x}Te_x$ (Lanthanide Dopants)	1.75 (775 K)	BM (15 min) + SPS (1073 K @ 60 Mpa)	Zhang et al. [84]
2019	$Mg_{3.2}Bi_{1.998-x}Sb_xTe_{0.002}$	0.9 (350 K)	BM (5 hrs) + HP (1053 K)	Mao et al. [23]
2019	$Mg_{3+\delta}Sb_xBi_{2-x}$ (Bi-Doping)	-	BM (10 hrs) + SPS (948 K @ 50 MPa)	Shu et al. [37]
2018	$Mg_{3.2}Sb_{1.5}Bi_{0.49}Te_{0.01}$ (Increasing Grain Size)	-	BM + SPS (873K /1123 K @ 50 MPa)	Kanno et al. [39]
2018	$Mn_xMg_{3.2-x}Sb_{1.5}Bi_{0.49}Te_{0.01}$	1.85 (723 K)	BM (10 hrs) + SPS (873K @50 MPa)	Chen et al. [91]
2017	$Mg_{3.2}Sb_{1.5}Bi_{0.49}Te_{0.01}$ Defect Engineering		BM + HP (923 -1073K) @ 80 MPa	Mao et al. [22]
2017	$Mg_{3.2}Sb_{1.5}Bi_{0.5-x}Te_x$(Te doping)	-	BM (10 hrs)-HP (923 K @ 50 MPa)	Mao et al. [35]
2017	$Mg_{3.2-x}Co_xSb_{1.5}Bi_{0.49}Te_{0.01}$	1.7 (773 K)	BM (10 hrs) + HP (923 K @ 80 MPa)	Mao et al. [92]
2017	$Mg_3Sb_{1.5-0.5x}Bi_{0.5-0.5x}Te_x$	1.65 (725 K)	BM (5 hrs)+ SPS (1123 K @ 75 MPa)	Zhang et al. [45]
2016	$Mg_{3+\delta}Sb_{2-x}Bi_{x-y}Te_y$	1.5 (716 K)	BM-SPS (873K @ 50 MPa)	Tamaki et al.[20]

*BM – Ball Milling; SPS – Spark Plasma Sintering; HP – Hot Pressing; SSR – Solid State Reaction

8. Phonon Dispersion and Thermal Properties

The pronounced potential of Mg_3Sb_2 and Mg_3Bi_2 emanates from their intrinsically low κ_L, typically measuring around 1-2 $Wm^{-1}K^{-1}$ at room temperature. This characteristic aligns with, and is comparable to, the κ of the established TE material Bi_2Te_3. The remarkably low κ_L observed in $Mg_3(Sb, Bi)_2$ challenges the conventional correlation between low κ_L and dense materials or compounds with complexity at the atomic or microstructural levels as shown in Figure 9(a). Despite its simple crystal structure and the lightweight Mg constituent element, Mg_3Sb_2 exhibits exceptionally high phonon scattering rates, low mean sound velocity (shown in Figure 9(b)) and surprisingly low κ_L. This anomaly is attributed to the soft and highly anharmonic transverse phonons, resulting in a large Grüneisen parameter, as revealed by phonon dispersion calculations using density functional theory (DFT) and high-temperature elasticity measurements. As indicated in Figure 9(c), Mg_3Sb_2 exhibits the presence of two distinct soft transverse acoustic (TA) modes. One is situated at the A point in the out-of-plane direction, while the other is located at the in-plane M point with a lower frequency.

The large Grüneisen parameter (~1.83 estimated for Mg_3Sb_2) displayed in Figure 9(d) is linked to the instability in the crystal structure, particularly at the octahedrally coordinated Mg(1) site, due to the smaller size of the Mg atoms. This leads to weak inter-layer bonding and soft shear modes, consistent with the smaller shear modulus values observed in Mg_3Sb_2 and Mg_3Bi_2 compared to other 1-2-2 Zintl materials as indicated in Figure 10. The inadequate fit of the Mg cation is believed to result in weak interlayer bonding, leading to observed soft shear modes and, ultimately, highly anharmonic behaviour in the acoustic branches. Experimental validation also includes the rapid decrease of the speed of sound and elastic moduli in Mg_3Sb_2 and Mg_3Bi_2 with increasing temperature, reflecting the softening of acoustic modes. The atom-projected phonon density of states analysis for Mg_3Sb_2 displayed in Figure 9(e), reveals that the low-, mid-, and high-frequency regimes are primarily governed by displacements of the anions (Sb), cations Mg(1), and the metal site (Mg(2)), respectively. Interestingly, the species on the cation site (Mg) does not exert a strong influence on the maximum frequency or the average phonon velocities.

The anharmonic characteristic, originating from the instability in the crystal structure, is consistent with the significantly smaller shear modulus values of Mg_3Sb_2 and Mg_3Bi_2 compared to other 1-2-2 Zintl materials. Moreover, these findings suggest a broader implication that soft shear modes arising from undersized cations offer a potential route to achieving low lattice κ_L in ionic layered structures. Elastic bulk (*K*) and shear (*G*) moduli of AMg_2X_2 compounds shown in Figure 10(a) and 10(b), as observed experimentally (asterisks) and computed (circles), exhibit a trend of decreasing values with increasing unit cell volume and bond length [48]. The notably low shear moduli of Mg_3Sb_2 and Mg_3Bi_2 are particularly significant, indicating a distinctive and soft bonding characteristic

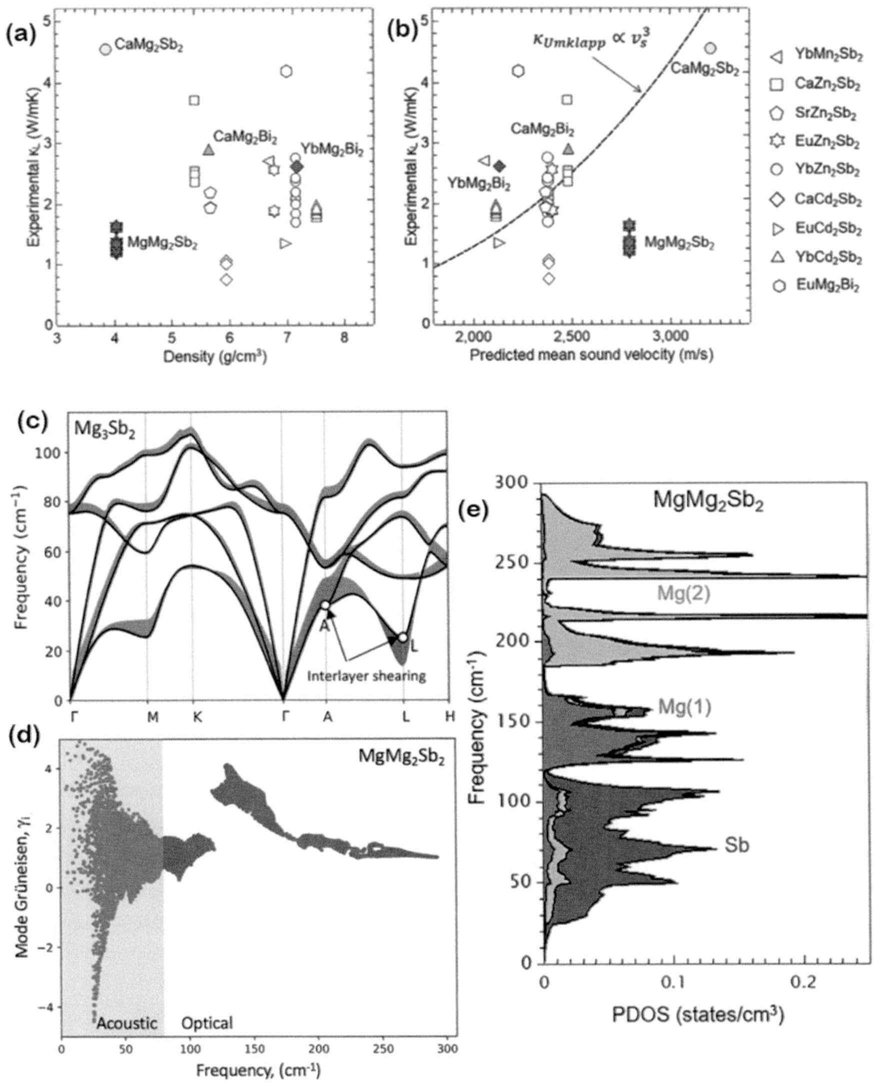

Figure 9. Comparison of Lattice Thermal Conductivity with (a) Density and (b) Mean sound velocity. indicating a significantly lower κ_L, of Mg_3Sb_2 among isostructural AM_2X_2 compounds. (c) The phonon dispersions of Mg_3Sb_2, in the low-frequency regime. (d) The mode Grüneisen parameters as a function of frequency. (e) Atom-projected phonon density of states of Mg_3Sb_2. (Reproduced with permission from Peng et al. [48] Copyright 2018, John Wiley & Sons, Inc.)

specific to these binary compounds [93]. Neutron scattering experiments support a trend of the entire phonon density of states becoming softer from Yb to Mg or from Ca to Mg in AMg_2Sb_2, contrary to the common trend of hardening with lighter elements. A comprehensive investigation [94] from first principles, having a consideration of anharmonicity, isotope, and boundary scattering processes and utilizing an accurate solution of the Boltzmann equation as shown in Figure evaluates the observed anomalously low κ to the grain-boundary scattering of phonons, primarily with purely anharmonic κ of perfect crystals to be an order of magnitude larger. Additionally, the contribution of mass disorder resulting from Bi-alloying and off-stoichiometry in nanopolycrystalline material led to further reduction in κ, showing good agreement with typical experiments [94].

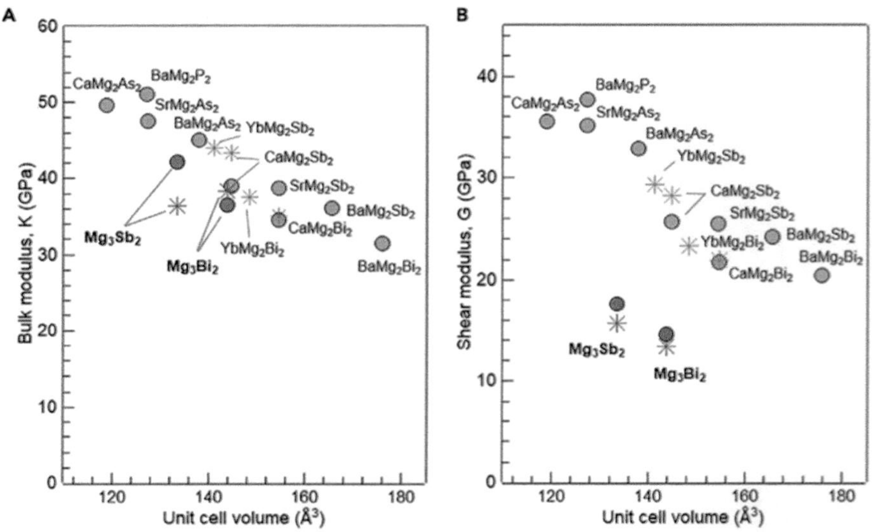

Figure 10. Comparison of Measured and Computed Elastic Moduli for AMg_2X_2 compounds with $CaAl_2Si_2$ type structure. (Reproduced with permission from Peng et al. [48] Copyright 2018, John Wiley & Sons, Inc.)

9. Thermoelectric Devices, Performance and Operational Stability

Emerging trends in TE materials, such as $Mg_3(Sb, Bi)_2$ Zintl phases, involve the exploration of novel approaches/compositions and combinations, as well as the potential for integration into large-scale energy harvesting systems. Efforts are underway to integrate Zintl TE materials into practical systems, addressing scalability and manufacturing ability as tabulated in Table 8, showing significant progress in the performance and potential of Mg_3Sb_2-based TE devices. Furthermore, the development of high-efficiency and reliable TE modules targets optimization of the resistivity of the barrier layer [95,96], besides attaining stable operation over extended periods in thermal cycling tests [97]. Despite these advancements, challenges related to the assembly and reliability of Mg_3Sb_2-based TE devices persist, where efforts are now aimed to enhance the interface stability of these devices to meet more efficient and reliable requirements for practical applications. Ongoing efforts are focused on addressing the associated challenges to realize their full potential in practical TE applications.

Table 8. Recent developments in Mg_3Sb_2 based prototype TE devices with selected *n*-type and *p*-type materials with power densities, and efficiency

n	*p*	Power density (Wcm^{-2})	Efficiency (%)	Ref.
$Mg_{3.1}Co_{0.1}Sb_{1.5}Bi_{0.49}Te_{0.01}$	$ZrCoBi_{0.65}Sb_{0.15}Sn_{0.20}$	-	10.6	Zhu et al. [98]
$Mg_{3.2}Sb1Bi_{0.996}Se_{0.004}$	$Mg_{1.98}ZnAg_{0.02}Sb_2$	0.23 (380 K)	7.3% (380 K)	Jiang et al. [97]
$Mg_{3.2}Sb_{1.5}Bi_{0.49}Te_{0.01}Cu_{0.01}$	α-MgAgSb	-	~7.3% (593 K)	Liu et al. [100]
$Mg_{3.2}Bi_{1.4975}Sb_{0.5}Te_{0.0025}$	$(Sb_{0.75}Bi_{0.25})_2(Te_{0.97}Se_{0.03})_3$	3.1 (575 K)	-	Yang et al. [101]
$Mg_{3.2}Sb_{1.5}Bi_{0.49}Te_{0.01}$	$Ge_{0.83}In_{0.01}Ti_{0.02}Sb_{0.09}Te$	1.94 (473 K)	10.1%	Ando et al. [102]
$Mg_{3.2}SbBi_{0.99}Te_{0.01}$	$Mg_{1.8-x}Na_xZn_{1.2}Sb_2$	-	5.5%	Liang et al. [103]
$Mg_{3.1}Co_{0.1}Sb_{1.5}Bi_{0.49}Te_{0.01}$	$ZrCoBi_{0.65}Sb_{0.15}Sn_{0.2}$	-	10.6% (~400K	Zhu et al. [98]
$Mg_{3.05}(Sb_{0.75}Bi_{0.25})_{1.99}Te_{0.01}(MgB_2)_{0.03}$	GeTe	-	5.7% (438 K)	Geng et al. [104]
$Mg_{3.15}Co_{0.05}Sb_{1.24}Bi_{0.75}Se_{0.01}$	$(Pb_{0.15}Ge_{0.85}Te)_{0.8}(AgSbTe_2)_{0.2}$	0.78 (400 K)	14.5% (480 K)	Li et al. [105]
$Fe/Mg_{3.2}Sb_{0.5}Bi_{1.495}Mn_{0.02}Te_{0.005}/Fe$ $Mg_{3.2}Sb_1Bi_{0.995}Mn_{0.02}Te_{0.005}/Fe/Ni$	$Ni/Li_{0.0025}(Bi_{0.4}Sb_{1.6})_{0.9975}Te_{3.01}/Ni/$ $Sb_{1.85}In_{0.15}Cu_{0.03}Te_{2.98}/Ni$	0.53 (400 K)	10.5% (380 K)	Sun et al. [96]
$Mg_{3.085}Y_{0.015}SbBi$	-	-	7	Shi et al. [69]
$Mg_{3.17}B_{0.03}Sb_{1.5}Bi_{0.49}Te_{0.0}$	$(Ca_{0.5}Yb_{0.25}Ba_{0.25})_{0.995}Na_{0.005}Mg_2Bi_{1.98}$	2.2 (655 K)	6.2	Guo et al. [99]

10. Concluding Remarks and Outlook

The exploration of $Mg_3(Sb, Bi)_2$ based Zintl phases for TE applications reveals promising avenues for sustainable energy conversion. The recent discovery of *n*-type $Mg_3(Sb, Bi)_2$ with peak zT higher than 1.5 and peak power factor around 20 mW $cm^{-1}K^{-2}$ has revived the interest in further improving the performance of the *p*-type counterpart. Despite its limitations, the Zintl formalism has remained a valuable tool for understanding the properties of $Mg_3(Sb, Bi)_2$ based TE compounds and other Zintl materials. The synthesis of $Mg_3(Sb, Bi)_2$ based TE materials presents processing challenges related to stability, scalability, and better control over their structural and electronic characteristics. Ongoing efforts aims to provide better understanding of site preference, bond length difference in anionic networks, achieving *p*-type conduction, controlling carrier concentration, regulating the effects of Mg content and Sb/Bi ratio, and exploring the microstructure-property relationship. A careful alloy design, precise synthesis techniques, and engineering at the nanoscale interfaces will enable an improved conversion efficiency. Challenges related to chemical stability under high-temperature and operating conditions desires further investigation into the pyrophoric nature of these compositions. The crystalline order, tunable electronic properties via aliovalent substitutions, and intrinsically low κ_L owing to their structural and chemical complexity motivate deeper investigations to propel $Mg_3(Sb, Bi)_2$-based thermoelectrics with enhanced conversion efficiencies into practical energy harvesting, solid-state-cooling, and power generation applications.

References

1. L.E. Bell, Cooling, heating, generating power, and recovering waste heat with thermoelectric systems, Science, 321 (2008) 1457-1461.
2. S. LeBlanc, Thermoelectric generators: Linking material properties and systems engineering for waste heat recovery applications, Sustainable Mater. Technol., 1-2 (2014) 26-35. https://doi.org/10.1016/j.susmat.2014.11.002.
3. D.M. Rowe, Applications of nuclear-powered thermoelectric generators in space, Applied Energy, 40 (1991) 241-271. https://doi.org/10.1016/0306-2619(91)90020-X.
4. C. Candolfi, S.E. Oualid, D. Ibrahim, S. Misra, O.E. Hamouli, A. Léon, A. Dauscher, P. Masschelein, P. Gall, P. Gougeon, C. Semprimoschnig, B. Lenoir, Thermoelectric materials for space applications, CEAS Space J., 13 (2021) 325-340. https://doi.org/10.1007/s12567-021-00351-x.
5. G.J. Snyder, E.S. Toberer, Complex thermoelectric materials, Nat. Mater., 7 (2008) 105-114. https://doi.org/10.1038/nmat2090.
6. C.B. Vining, An inconvenient truth about thermoelectrics, Nat. Mater., 8 (2009) 83-85.

7. N. Kazem, S.M. Kauzlarich, Thermoelectric properties of Zintl antimonides. pp. 177-208. *In:* Handbook on the Physics and Chemistry of Rare Earths, Elsevier (2016).
8. R. Nesper, The Zintl-Klemm concept – A historical survey, Z. Anorg. Allg. Chem., 640 (2014) 2639-2648. https://doi.org/10.1002/zaac.201400403.
9. S.-M. Park, S.-J. Kim, Sr11Cd6Sb12: A new Zintl compound with infinite chains of pentagonal tubes, J. Solid State Chem., 177 (2004) 3418-3422.
10. B. Saparov, S. Bobev, A. Ozbay, E.R. Nowak, Synthesis, structure and physical properties of the new Zintl phases $Eu_{11}Zn_6Sb_{12}$ and $Eu_{11}Cd_6Sb_{12}$, J. Solid State Chem., 181 (2008) 2690-2696.
11. S.K. Bux, A. Zevalkink, O. Janka, D. Uhl, S. Kauzlarich, J.G. Snyder, J.-P. Fleurial, Glass-like lattice thermal conductivity and high thermoelectric efficiency in $Yb_9Mn_{4.2}Sb_9$, J. Mater. Chem. A, 2 (2014) 215-220.
12. S. Ohno, A. Zevalkink, Y. Takagiwa, S.K. Bux, G.J. Snyder, Thermoelectric properties of the $Yb_9Mn_{4.2-x}Zn_xSb_9$ solid solutions, J. Mater. Chem. A, 2 (2014) 7478-7483.
13. F. Gascoin, S. Ottensmann, D. Stark, S.M. Haïle, G.J. Snyder, Zintl phases as thermoelectric materials: Tuned transport properties of the compounds $CaxYb_{1-x}Zn_2Sb_2$, Adv. Funct. Mater., 15 (2005) 1860-1864.
14. O.Y. Zelinska, A.V. Tkachuk, A.P. Grosvenor, A. Mar, Structure and physical properties of $YbZn_2Sb_2$ and $YbCd_2Sb_2$, Chem. Met. Alloys (2008) 204-209.
15. A. Morozkin, O. Isnard, P. Henry, S. Granovsky, R. Nirmala, P. Manfrinetti, Synthesis and magnetic structure of the $YbMn_2Sb_2$ compound, J. Alloys Compd., 420 (2006) 34-36.
16. T. Zhu, C. Yu, J. He, S. Zhang, X. Zhao, T.M. Tritt, Thermoelectric Properties of Zintl Compound $YbZn_2Sb_2$ with Mn Substitution in Anionic Framework, J. Electron. Mater., 38 (2009) 1068-1071.
17. H. Zhang, J.-T. Zhao, Y. Grin, X.-J. Wang, M.-B. Tang, Z.-Y. Man, H.-H. Chen, X.-X. Yang, A new type of thermoelectric material, $EuZn_2Sb_2$, J. Chem. Phys., 129 (2008).
18. J.H. Grebenkemper, Y. Hu, D. Barrett, P. Gogna, C.-K. Huang, S.K. Bux, S.M. Kauzlarich, High temperature thermoelectric properties of $Yb_{14}MnSb_{11}$ prepared from reaction of MnSb with the elements, Chem. Mater., 27 (2015) 5791-5798. https://doi.org/10.1021/acs.chemmater.5b02446.
19. S.R. Brown, S.M. Kauzlarich, F. Gascoin, G.J. Snyder, $Yb_{14}MnSb_{11}$: New high efficiency thermoelectric material for power generation, Chem. Mater., 18 (2006) 1873-1877.
20. H. Tamaki, H.K. Sato, T. Kanno, Isotropic conduction network and defect chemistry in Mg^{3+} δSb2-based layered zintl compounds with high thermoelectric performance, Adv. Mater., 28 (2016) 10182-10187.
21. J. Shuai, J. Mao, S. Song, Q. Zhu, J. Sun, Y. Wang, R. He, J. Zhou, G. Chen, D.J. Singh, Tuning the carrier scattering mechanism to effectively improve the thermoelectric properties, Energy Environ. Sci. 10 (2017) 799-807.
22. J. Mao, Y. Wu, S. Song, Q. Zhu, J. Shuai, Z. Liu, Y. Pei, Z. Ren, Defect engineering for realizing high thermoelectric performance in *n*-type Mg_3Sb_2-based materials, ACS Energy Lett., 2 (2017) 2245-2250. https://doi.org/10.1021/acsenergylett.7b00742.
23. J. Mao, H. Zhu, Z. Ding, Z. Liu, G.A. Gamage, G. Chen, Z. Ren, High thermoelectric cooling performance of *n*-type Mg_3Bi_2-based materials, Science, 365 (2019) 495-498. https://doi.org/10.1126/science.aax7792.

24. S.M. Kauzlarich, Special issue: Advances in zintl phases, Materials, 12 (2019) 2554. https://doi.org/10.3390/ma12162554.
25. S. Fang, J. Li, K. Zou, H. Shuai, L. Xu, W. Deng, G. Zou, H. Hou, X. Ji, Zintl chemistry: Current status and future perspectives, J. Chem. Eng., 433 (2022) 133841. https://doi.org/10.1016/j.cej.2021.133841.
26. J. Shuai, J. Mao, S. Song, Q. Zhang, G. Chen, Z. Ren, Recent progress and future challenges on thermoelectric Zintl materials, Mater. Today Phys., 1 (2017) 74-95. https://doi.org/10.1016/j.mtphys.2017.06.003.
27. S.M. Kauzlarich, S.R. Brown, G. Jeffrey Snyder, Zintl phases for thermoelectric devices, Dalton Trans. (2007) 2099. https://doi.org/10.1039/b702266b.
28. G.A. Papoian, R. Hoffmann, Hypervalent bonding in one, two, and three dimensions: Extending the Zintl–Klemm concept to nonclassical electron-rich networks, Angew. Chem., Int. Ed., 39 (2000) 2408-2448.
29. S. Chanakian, U. Aydemir, A. Zevalkink, Z.M. Gibbs, J.-P. Fleurial, S. Bux, G.J. Snyder, High temperature thermoelectric properties of Zn-doped $Eu_5In_2Sb_6$, J. Mater. Chem. C. 3 (2015) 10518-10524.
30. J. Shuai, H. Geng, Y. Lan, Z. Zhu, C. Wang, Z. Liu, J. Bao, C.W. Chu, J. Sui, Z. Ren, Higher thermoelectric performance of Zintl phases $(Eu_{0.5}Yb_{0.5})_{1-x}Ca_xMg_2Bi_2$ by band engineering and strain fluctuation, Proceed. Nat. Acad. Sci. USA, 113 (2016) E4125-E4132. https://doi.org/10.1073/pnas.1608794113.
31. S. Ohno, K. Imasato, S. Anand, H. Tamaki, S.D. Kang, P. Gorai, H.K. Sato, E.S. Toberer, T. Kanno, G.J. Snyder, Phase boundary mapping to obtain *n*-type Mg_3Sb_2-based thermoelectrics, Joule, 2 (2018) 141-154. https://doi.org/10.1016/j.joule.2017.11.005.
32. F. Hao, P. Qiu, Y. Tang, S. Bai, T. Xing, H.S. Chu, Q. Zhang, P. Lu, T. Zhang, D. Ren, J. Chen, X. Shi, L. Chen, High efficiency Bi_2Te_3-based materials and devices for thermoelectric power generation between 100 and 300 °C, Energy Environ. Sci., 9 (2016) 3120-3127. https://doi.org/10.1039/c6ee02017h.
33. M. Wood, J.J. Kuo, K. Imasato, G.J. Snyder, Improvement of low-temperature zT in a Mg_3 Sb_2-Mg_3 Bi_2 solid solution via Mg-vapor annealing, Adv. Mater., 31 (2019) 1902337. https://doi.org/10.1002/adma.201902337.
34. M. Wood, K. Imasato, S. Anand, J. Yang, G.J. Snyder, The importance of the Mg-Mg interaction in Mg_3Sb_2-Mg_3Bi_2 shown through cation site alloying, J. Mater. Chem. A, 8 (2020) 2033-2038. https://doi.org/10.1039/C9TA11328B.
35. J. Mao, Y. Wu, S. Song, J. Shuai, Z. Liu, Y. Pei, Z. Ren, Anomalous electrical conductivity of n-type Te-doped $Mg_{3.2}Sb_{1.5}Bi_{0.5}$, Mater. Today Phys. 3 (2017) 1-6.
36. X. Shi, C. Sun, Z. Bu, X. Zhang, Y. Wu, S. Lin, W. Li, A. Faghaninia, A. Jain, Y. Pei, Revelation of inherently high mobility enables Mg_3Sb_2 as a sustainable alternative to n-Bi_2Te_3 thermoelectrics, Adv. Sci., 6 (2019) 1802286. https://doi.org/10.1002/advs.201802286.
37. R. Shu, Y. Zhou, Q. Wang, Z. Han, Y. Zhu, Y. Liu, Y. Chen, M. Gu, W. Xu, Y. Wang, W. Zhang, L. Huang, W. Liu, Thermoelectrics: Mg^{3+} δSb_x $Bi^{2-}{}_x$ family: A promising substitute for the state-of-the-art n-type thermoelectric materials near room temperature, Adv. Funct. Mater., 29 (2019) 1970020. https://doi.org/10.1002/adfm.201970020.
38. J. Li, S. Zhang, S. Zheng, Z. Zhang, B. Wang, L. Chen, G. Lu, Defect chemistry for n-type doping of Mg_3 Sb_2-based thermoelectric materials, J. Phys. Chem. C, 123 (2019) 20781-20788. https://doi.org/10.1021/acs.jpcc.9b05859.

39. T. Kanno, H. Tamaki, H.K. Sato, S.D. Kang, S. Ohno, K. Imasato, J.J. Kuo, G.J. Snyder, Y. Miyazaki, Enhancement of average thermoelectric figure of merit by increasing the grain-size of $Mg_{3.2}Sb_{1.5}Bi_{0.49}Te_{0.01}$, Appl. Phys. Lett., 112 (2018).
40. T. Luo, J.J. Kuo, K.J. Griffith, K. Imasato, O. Cojocaru-Mirédin, M. Wuttig, B. Gault, Y. Yu, G.J. Snyder, Nb-mediated grain growth and grain-boundary engineering in Mg_3Sb_2-based thermoelectric materials, Adv. Funct. Mater., 31 (2021) 2100258. https://doi.org/10.1002/adfm.202100258.
41. K. Imasato, M. Wood, S. Anand, J.J. Kuo, G.J. Snyder, Understanding the high thermoelectric performance of Mg_3Sb_2-Mg_3Bi_2 alloys, Adv. Energy and Sustain. Res., 3 (2022) 2100208. https://doi.org/10.1002/aesr.202100208.
42. E. Zintl, E. Husemann, Bindungsart und Gitterbau binärer Magnesiumverbindungen, Z. Phys. Chem. 21B (1933) 138-155. https://doi.org/10.1515/zpch-1933-2112.
43. C. Zheng, R. Hoffmann, R. Nesper, H.G. Von Schnering, Site preferences and bond length differences in $CaAl_2Si_2$-type Zintl compounds, J. Am. Chem., 108 (1986) 1876-1884.
44. A.A. Nayeb-Hashemi, J.B. Clark, The Bi-Mg (Bismuth-Magnesium) system, Bull. Alloy Phase Diagrams, 6 (1985) 528-533. https://doi.org/10.1007/BF02887150.
45. J. Zhang, L. Song, S.H. Pedersen, H. Yin, L.T. Hung, B.B. Iversen, Discovery of high-performance low-cost n-type Mg_3Sb_2-based thermoelectric materials with multi-valley conduction bands, Nat. Commun., 8 (2017) 13901.
46. J. Wang, J. Mark, K.E. Woo, J. Voyles, K. Kovnir, Chemical flexibility of Mg in pnictide materials: Structure and properties diversity, Chem. Mater., 31 (2019) 8286-8300.
47. X. Sun, X. Li, J. Yang, J. Xi, R. Nelson, C. Ertural, R. Dronskowski, W. Liu, G.J. Snyder, D.J. Singh, W. Zhang, Achieving band convergence by tuning the bonding ionicity in *n*-type Mg_3Sb_2, J. Comput. Chem., 40 (2019) 1693-1700. https://doi.org/10.1002/jcc.25822.
48. W. Peng, G. Petretto, G.-M. Rignanese, G. Hautier, A. Zevalkink, An unlikely route to low lattice thermal conductivity: Small atoms in a simple layered structure, Joule, 2 (2018) 1879-1893.
49. R. Sanderson, Chemical Bonds and Bonds Energy, Elsevier (2012).
50. H. Okamoto, Mg-Sb (Magnesium-Antimony), J. Phase Equilib. Diffus., 31 (2010) 574-574.
51. X. Li, P. Yang, Y. Wang, Z. Zhang, D. Qin, W. Xue, C. Chen, Y. Huang, X. Xie, X. Wang, Phase boundary mapping in ZrNiSn half-Heusler for enhanced thermoelectric performance, Research, 2020 (2020).
52. M. Wood, M.Y. Toriyama, S. Dugar, J. Male, S. Anand, V. Stevanović, G.J. Snyder, Phase boundary mapping of tin-doped ZnSb reveals thermodynamic route to high thermoelectric efficiency, Adv. Energy Mater., 11 (2021) 1-8. https://doi.org/10.1002/aenm.202100181.
53. J.H. Slowik, Core excitons in amorphous magnesium alloys, Phys. Rev. B., 10 (1974) 416-431. https://doi.org/10.1103/PhysRevB.10.416.
54. L.M. Watson, C.A.W. Marshall, C.P. Cardoso, On the electronic structure of the semiconducting compounds Mg_3 Bi_2 and Mg_3 Sb_2, J. Phys. F: Met. Phys., 14 (1984) 113-121. https://doi.org/10.1088/0305-4608/14/1/014.
55. T. Kajikawa, N. Kimura, T. Yokoyama, Thermoelectric properties of intermetallic compounds: Mg_3Bi_2 and Mg_3Sb_2 for medium temperature range thermoelectric elements, *In:* Proceedings ICT'03. 22nd International Conference on Thermoelectrics

(IEEE Cat. No.03TH8726), IEEE, La Grande Motte, France, 2003: pp. 305-308. https://doi.org/10.1109/ICT.2003.1287510.
56. V. Ponnambalam, D.T. Morelli, On the thermoelectric properties of Zintl compounds $Mg_3Bi_{2-x}Pn_x$ (Pn = P and Sb), J. Electron. Mater., 42 (2013) 1307-1312. https://doi.org/10.1007/s11664-012-2417-7.
57. J. Tani, H. Ishikawa, One-step rapid synthesis of *n*-type Y-doped Mg_3Sb_2 by pulsed electric current sintering and investigation of its thermoelectric properties, Mater. Lett. 262 (2020) 127056. https://doi.org/10.1016/j.matlet.2019.127056.
58. J. Tani, H. Ishikawa, Thermoelectric properties of Te-doped Mg_3Sb_2 synthesized by spark plasma sintering, Phys. B (Amsterdam, Neth.), 588 (2020) 412173. https://doi.org/10.1016/j.physb.2020.412173.
59. R. Bhardwaj, A. Bhattacharya, K. Tyagi, B. Gahtori, N.S. Chauhan, A. Vishwakarma, K.K. Johari, S. Bathula, S. Auluck, A. Dhar, Enhancement in thermoelectric performance of single step synthesized Mg doped Cu_2Se: An experimental and theoretical study, Intermetallics, 112 (2019) 106541. https://doi.org/10.1016/j.intermet.2019.106541.
60. S.H. Kim, C.M. Kim, Y.-K. Hong, K.I. Sim, J.H. Kim, T. Onimaru, T. Takabatake, M.-H. Jung, Thermoelectric properties of $Mg_3Sb_{2-x}Bi_x$ single crystals grown by Bridgman method, Mater. Res. Express., 2 (2015) 055903. https://doi.org/10.1088/2053-1591/2/5/055903.
61. K. Imasato, C. Fu, Y. Pan, M. Wood, J.J. Kuo, C. Felser, G.J. Snyder, Metallic *n*-type Mg_3Sb_2 single crystals demonstrate the absence of ionized impurity scattering and enhanced thermoelectric performance, Adv. Mater., 32 (2020) 1908218. https://doi.org/10.1002/adma.201908218.
62. M. Jin, S. Lin, W. Li, X. Zhang, Y. Pei, Nearly isotropic transport properties in anisotropically structured n-type single-crystalline Mg_3Sb_2, Mater. Today Phys., 21 (2021) 100508. https://doi.org/10.1016/j.mtphys.2021.100508.
63. F. Ahmadpour, T. Kolodiazhnyi, Y. Mozharivskyj, Structural and physical properties of $Mg_{3-x}Zn_xSb_2$ (x=0-1.34), J. Solid State Chem., 180 (2007) 2420-2428. https://doi.org/10.1016/j.jssc.2007.06.011.
64. Z. Ren, J. Shuai, J. Mao, Q. Zhu, S. Song, Y. Ni, S. Chen, Significantly enhanced thermoelectric properties of *p*-type Mg_3Sb_2 via co-doping of Na and Zn, Acta Materialia, 143 (2018) 265-271. https://doi.org/10.1016/j.actamat.2017.10.015.
65. C.L. Condron, S.M. Kauzlarich, F. Gascoin, G.J. Snyder, Thermoelectric properties and microstructure of Mg_3Sb_2, J. Solid State Chem., 179 (2006) 2252-2257. https://doi.org/10.1016/j.jssc.2006.01.034.
66. J. Shuai, Y. Wang, H.S. Kim, Z. Liu, J. Sun, S. Chen, J. Sui, Z. Ren, Thermoelectric properties of Na-doped Zintl compound: $Mg_{3-x}Na_xSb_2$, Acta Materialia, 93 (2015) 187-193.
67. L. Song, J. Zhang, B.B. Iversen, Simultaneous improvement of power factor and thermal conductivity via Ag doping in *p*-type Mg_3Sb_2 thermoelectric materials, J. Mater. Chem. A, 5 (2017) 4932-4939. https://doi.org/10.1039/C6TA08316A.
68. A. Bhardwaj, N.S. Chauhan, S. Goel, V. Singh, J. Pulikkotil, T. Senguttuvan, D. Misra, Tuning the carrier concentration using Zintl chemistry in Mg_3Sb_2, and its implications for thermoelectric figure-of-merit, Phys. Chem. Chem. Phys., 18 (2016) 6191-6200.
69. X. Shi, X. Zhang, A. Ganose, J. Park, C. Sun, Z. Chen, S. Lin, W. Li, A. Jain, Y. Pei, Compromise between band structure and phonon scattering in efficient n-Mg_3Sb_2-Bi

thermoelectrics, Mater. Today Phys., 18 (2021) 100362. https://doi.org/10.1016/j.mtphys.2021.100362.

70. D.W. Zhou, J.S. Liu, S.H. Xu, P. Peng, Thermal stability and elastic properties of Mg_3Sb_2 and Mg_3Bi_2 phases from first-principles calculations, Phys. B (Amsterdam, Neth.), 405 (2010) 2863-2868. https://doi.org/10.1016/j.physb.2010.04.013.
71. H.X. Xin, X.Y. Qin, Electrical and thermoelectric properties of nanocrystal substitutional semiconductor alloys $Mg_3(Bi_xSb_{1-x})_2$ prepared by mechanical alloying, J. Phys. D: Appl. Phys., 39 (2006) 5331-5337. https://doi.org/10.1088/0022-3727/39/24/035.
72. A. Bhardwaj, A.K. Shukla, S.R. Dhakate, D.K. Misra, Graphene boosts thermoelectric performance of a Zintl phase compound, RSC Adv., 5 (2015) 11058-11070. https://doi.org/10.1039/c4ra15456h.
73. A. Bhardwaj, N.S. Chauhan, D. Misra, Significantly enhanced thermoelectric figure of merit of *p*-type Mg_3Sb_2-based Zintl phase compounds via nanostructuring and employing high energy mechanical milling coupled with spark plasma sintering, J. Mater. Chem. A, 3 (2015) 10777-10786.
74. C. Chen, X. Li, S. Li, X. Wang, Z. Zhang, J. Sui, F. Cao, X. Liu, Q. Zhang, Enhanced thermoelectric performance of *p*-type Mg_3Sb_2 by lithium doping and its tunability in an anionic framework, J. Mater. Sci., 53 (2018) 16001-16009. https://doi.org/10.1007/s10853-018-2555-2.
75. X. Tang, B. Zhang, X. Zhang, S. Wang, X. Lu, G. Han, G. Wang, X. Zhou, Enhancing the thermoelectric performance of *p*-type Mg_3Sb_2 via codoping of Li and Cd, ACS Appl. Mater. Interfaces. 12 (2020) 8359-8365. https://doi.org/10.1021/acsami.9b23059.
76. Y. Cui, X. Zhang, B. Duan, J. Li, H. Yang, H. Wang, P. Wen, T. Gao, Z. Fang, G. Li, Y. Li, P. Zhai, Band structure and thermoelectric properties of Al-doped $Mg_{3-x}Al_xSb_2$ compounds, J. Mater. Sci.: Mater. Electron., 30 (2019) 15206-15213. https://doi.org/10.1007/s10854-019-01893-x.
77. X. Li, H. Xie, B. Yang, Anisotropic electronic transport properties of Ag-doped Mg_3Sb_2 crystal prepared by directional solidification, J. Appl. Phys. (Melville, NY, U.S.). 127 (2020) 195104. https://doi.org/10.1063/5.0006340.
78. Y. Fu, X. Zhang, H. Liu, J. Tian, J. Zhang, Thermoelectric properties of Ag-doped compound: $Mg_{3-x}Ag_xSb_2$, J. Materiomics, 4 (2018) 75-79. https://doi.org/10.1016/j.jmat.2017.12.002.
79. J. Zhang, L. Song, M. Sist, K. Tolborg, B.B. Iversen, Chemical bonding origin of the unexpected isotropic physical properties in thermoelectric Mg_3Sb_2 and related materials, Nat. Commun., 9 (2018) 4716. https://doi.org/10.1038/s41467-018-06980-x.
80. F. Meng, S. Sun, J. Ma, C. Chronister, J. He, W. Li, Anisotropic thermoelectric figure-of-merit in Mg_3Sb_2, Mater. Today Phys., 13 (2020) 100217. https://doi.org/10.1016/j.mtphys.2020.100217.
81. A. Bhardwaj, D. Misra, Enhancing thermoelectric properties of a *p*-type Mg_3Sb_2-based Zintl phase compound by Pb substitution in the anionic framework, RSC Adv., 4 (2014) 34552-34560.
82. J.J. Kuo, Y. Yu, S.D. Kang, O. Cojocaru-Mirédin, M. Wuttig, G.J. Snyder, Mg deficiency in grain boundaries of *n*-type Mg_3Sb_2 identified by atom probe tomography, Adv. Mater. Interfaces., 6 (2019) 1900429. https://doi.org/10.1002/admi.201900429.
83. L. Yu, W. Li, Z. Zhang, S. Wei, J. Li, Z. Ji, J. Zhuo, G. Lu, W. Song, S. Zheng,

Significantly enhanced thermoelectric figure of merit of *n*-type Mg_3Sb_2-based Zintl phase compounds via co-doped of Mg and Sb site, Mater. Today Phys., 26 (2022) 100721. https://doi.org/10.1016/j.mtphys.2022.100721.

84. J. Zhang, L. Song, B.B. Iversen, Probing efficient *n*-type lanthanide dopants for Mg_3Sb_2 thermoelectrics, Adv. Sci., 7 (2020) 2002867. https://doi.org/10.1002/advs.202002867.
85. J.-W. Li, Z. Han, J. Yu, H.-L. Zhuang, H. Hu, B. Su, H. Li, Y. Jiang, L. Chen, W. Liu, Wide-temperature-range thermoelectric *n*-type $Mg_3(Sb, Bi)_2$ with high average and peak zT values, Nat. Commun., 14 (2023) 7428.
86. Z. Li, C. Sun, X. Li, X. Ye, K. Yang, X. Nie, W. Zhao, Q. Zhang, Realizing excellent structural and thermoelectric performance in Mg_3Sb_2-based alloys by manipulating Mg intrinsic migration kinetics with interstitial Ni doping, ACS Appl. Mater. Interfaces. (2023) acsami.3c02351. https://doi.org/10.1021/acsami.3c02351.
87. Y. Zhu, R. Shu, F. Jiang, A. Elsukova, P.O.Å. Persson, P. Eklund, W. Liu, Enhanced thermoelectric properties of Mg_2Sn-Mg_3Sb_2 nanocomposites by tailoring matrix/inclusion interface toward energy harvesting applications, ACS Appl. Nano Mater., 6 (2023) 6133-6140. https://doi.org/10.1021/acsanm.3c00439.
88. J. Liang, X. Shi, Y. Peng, W. Liu, H. Yang, C. Liu, J. Chen, Q. Zhou, L. Miao, Z. Chen, Synergistic effect of band and nanostructure engineering on the boosted thermoelectric performance of *n*-type $Mg^{3+\delta}$ $(Sb, Bi)_2$ Zintls, Adv. Energy Mater., 12 (2022) 2201086.
89. R. Shu, Z. Han, A. Elsukova, Y. Zhu, P. Qin, F. Jiang, J. Lu, P.O. Persson, J. Palisaitis, A. Le Febvrier, Solid-state janus nanoprecipitation enables amorphous-like heat conduction in crystalline Mg_3Sb_2-based thermoelectric materials, Adv. Sci., 9 (2022) 2202594.
90. J. Zhang, L. Song, B.B. Iversen, Improved thermoelectric properties of N-type Mg_3Sb_2 through cation-site doping with Gd or Ho, ACS Appl. Mater. Interfaces, 13 (2021) 10964-10971. https://doi.org/10.1021/acsami.0c22558.
91. X. Chen, H. Wu, J. Cui, Y. Xiao, Y. Zhang, J. He, Y. Chen, J. Cao, W. Cai, S.J. Pennycook, Z. Liu, L.-D. Zhao, J. Sui, Extraordinary thermoelectric performance in *n*-type manganese doped Mg_3Sb_2 Zintl: High band degeneracy, tuned carrier scattering mechanism and hierarchical microstructure, Nano Energy, 52 (2018) 246-255. https://doi.org/10.1016/j.nanoen.2018.07.059.
92. J. Mao, J. Shuai, S. Song, Y. Wu, R. Dally, J. Zhou, Z. Liu, J. Sun, Q. Zhang, C. Dela Cruz, S. Wilson, Y. Pei, D.J. Singh, G. Chen, C.-W. Chu, Z. Ren, Manipulation of ionized impurity scattering for achieving high thermoelectric performance in n-type Mg_3Sb_2-based materials, Proc. Natl. Acad. Sci. U.S.A. 114 (2017) 10548-10553. https://doi.org/10.1073/pnas.1711725114.
93. M. De Jong, W. Chen, T. Angsten, A. Jain, R. Notestine, A. Gamst, M. Sluiter, C. Krishna Ande, S. Van Der Zwaag, J.J. Plata, C. Toher, S. Curtarolo, G. Ceder, K.A. Persson, M. Asta, Charting the complete elastic properties of inorganic crystalline compounds, Sci. Data, 2 (2015) 150009. https://doi.org/10.1038/sdata.2015.9.
94. M.B. Maccioni, R. Farris, V. Fiorentini, *Ab initio* thermal conductivity of thermoelectric Mg_3Sb_2: Evidence for dominant extrinsic effects, Phys. Rev. B, 98 (2018) 220301. https://doi.org/10.1103/PhysRevB.98.220301.
95. H. Zhang, Y. Wang, Z.A. Munir, Y. Zhang, W. Fan, S. Chen, Improvement of the conversion efficiency of Mg_3Sb_2 thermoelectric devices through optimizing the resistivity of the MgSbNi barrier layer, J. Mater. Sci. Technol. (Shenyang, China), 168 (2024) 208-214. https://doi.org/10.1016/j.jmst.2023.05.034.

96. Y. Sun, Y. Zhu, H. Wu, N. Qu, L. Xie, J. Zhu, Z. Liu, Q. Zhang, W. Cai, F. Guo, J. Sui, Rational design from materials to devices enables an efficiency of 10.5% based on thermoelectric $(Bi, Sb)_2$ Te_3 and Mg_3 $(Bi, Sb)_2$ for power generation, Energy Environ. Sci. (2024) 10.1039.D3EE03411A. https://doi.org/10.1039/D3EE03411A.
97. M. Jiang, Y. Fu, Q. Zhang, Z. Hu, A. Huang, S. Wang, L. Wang, W. Jiang, High-efficiency and reliable same-parent thermoelectric modules using Mg_3Sb_2-based compounds, National Sci. Rev., 10 (2023) nwad095. https://doi.org/10.1093/nsr/nwad095.
98. Q. Zhu, S. Song, H. Zhu, Z. Ren, Realizing high conversion efficiency of Mg_3Sb_2-based thermoelectric materials, J. Power Sources, 414 (2019) 393-400. https://doi.org/10.1016/j.jpowsour.2019.01.022.
99. M. Guo, W. Zhai, J. Li, J. Zhu, F. Guo, Z. Liu, M. Liu, Y. Zhu, X. Dong, Y. Zhang, Q. Zhang, W. Cai, J. Sui, High thermoelectric performance of $CaMg_2$ Bi_2 enabled by dynamic doping and orbital alignment, Adv. Funct. Materials, 32 (2022) 2200407. https://doi.org/10.1002/adfm.202200407.
100. Z. Liu, N. Sato, W. Gao, K. Yubuta, N. Kawamoto, M. Mitome, K. Kurashima, Y. Owada, K. Nagase, C.-H. Lee, J. Yi, K. Tsuchiya, T. Mori, Demonstration of ultrahigh thermoelectric efficiency of ~7.3% in Mg_3Sb_2/MgAgSb module for low-temperature energy harvesting, Joule, 5 (2021) 1196-1208. https://doi.org/10.1016/j.joule.2021.03.017.
101. J. Yang, G. Li, H. Zhu, N. Chen, T. Lu, J. Gao, L. Guo, J. Xiang, P. Sun, Y. Yao, R. Yang, H. Zhao, Next-generation thermoelectric cooling modules based on high-performance Mg3(Bi,Sb)2 material, Joule, 6 (2022) 193-204. https://doi.org/10.1016/j.joule.2021.11.008.
102. F. Ando, H. Tamaki, Y. Matsumura, T. Urata, T. Kawabe, R. Yamamura, Y. Kaneko, R. Funahashi, T. Kanno, Dual-boost thermoelectric power generation in a GeTe/Mg_3Sb_2-based module, Mater. Today Phys., 36 (2023) 101156. https://doi.org/10.1016/j.mtphys.2023.101156.
103. Z. Liang, C. Xu, S. Song, X. Shi, W. Ren, Z. Ren, Enhanced thermoelectric performance of *p*-type Mg_3Sb_2 for reliable and low-cost all-Mg_3Sb_2-based thermoelectric low-grade heat recovery, Adv. Funct. Materials, 33 (2023) 2210016. https://doi.org/10.1002/adfm.202210016.
104. Y. Geng, Z. Li, Z. Lin, Y. Liu, Q. Lai, X. Wu, L. Hu, F. Liu, Y. Yu, C. Zhang, Inhibiting Mg diffusion and evaporation by forming Mg-rich reservoir at grain boundaries improves the thermal stability of *n*-type Mg_3Sb_2 thermoelectrics, Small, (2023) 2305670. https://doi.org/10.1002/smll.202305670.
105. X. Li, C. Chen, L. Yin, X. Wang, J. Mao, F. Cao, Q. Zhang, Realizing an excellent conversion efficiency of 14.5% in the Mg_3Sb_2/GeTe-based thermoelectric module for waste heat recovery, Energy Environ. Sci. 16 (2023) 6147-6154. https://doi.org/10.1039/D3EE02587J.

CHAPTER

3

The Role of Defect Rich Catalysts in Electrochemical Water Splitting

Komal Patil[1], Pravin Babar[2], Ruturaj Jadhav[3], Jin Hyeok Kim[3] and Jongsung Park[1*]

[1] Department of Energy Engineering, Gyeongsang National University, Jinju, Gyeongnam 52828, Republic of Korea

[2] Colorado State University, Fort Collins 80523, Colorado, USA

[3] Optoelectronic Convergence Research Center, Department of Materials Science and Engineering, Chonnam National University, Gwangju 500-757, South Korea

1. Introduction

Hydrogen serves as an eco-friendly and adaptable energy transporter, offering a pathway towards a carbon-neutral society [1,2]. Water electrolysis stands out as the most environmentally friendly method to generate pure hydrogen, presenting a viable approach to store renewable energy and promote sustainability in our energy ecosystem [3–5]. The process of water splitting into hydrogen and oxygen involves two half-reactions: the hydrogen evolution reaction (HER) occurring at the cathode and the oxygen evolution reaction (OER) at the anode [6–8]. In theory, it requires a voltage of 1.23 V to drive these electrochemical water-splitting reactions. However, practical implementation necessitates a higher over-potential (η) due to slow reaction kinetics, especially for the OER. Enhancing the efficiency of these reactions and reducing the over potential can be achieved through the development of highly effective catalysts [9,10].

Currently, the catalysts used in industrial water splitting are predominantly composed of precious metals. Platinum-based catalysts are used for the HER, while iridium or ruthenium-based catalysts are employed for the OER [11–13]. Nevertheless, the widespread use of these catalysts is hindered by their high cost due to the precious metals involved [14,15]. As a result, extensive research efforts have been directed towards creating cost-effective and efficient catalysts

*Corresponding author: j.park@gnu.ac.kr

[16,17]. This includes strategies like minimizing the use of precious metals or utilizing non-precious metal-based compounds such as perovskites [18–20], spinels [21], other oxides [22], and (oxy)hydroxides [23,24] for the OER, as well as chalcogenides (e.g., MoS_2) [25–27], phosphides [28], nitrides, and transition metal oxides for the HER.

The mass activity, which measures the current density per unit mass of the catalyst, plays a pivotal role in determining the real-world utility of catalysts. It hinges on two fundamental aspects: the quantity of active sites and the inherent reactivity of each of these sites. To amplify the number of active sites, nanostructure engineering comes into play, where catalysts are fine-tuned in terms of their size, shape, and porosity [29–31]. Concurrently, strategies encompassing phase modulation, compositional adjustments, orientation control, strain management, defect introduction, and similar techniques serve to elevate the intrinsic reactivity of these sites. Together, these methods enable the optimization of catalysts for practical applications by increasing their catalytic efficiency and effectiveness. Among all strategies, the defective engineering offers a versatile and powerful method to fine-tune the properties of catalyst materials, ultimately leading to improved electrocatalytic performance, increased energy efficiency, and the development of more sustainable and effective catalysts for applications such as water splitting [32–34].

This chapter provides an overview of recent advancements in defect engineering within the context of electrocatalysts for water splitting. We examine the multifaceted approaches taken to comprehend the impact of defects, employing both experimental and computational methods. Our exploration begins with a classification of various types of defects, followed by a comprehensive discussion of the defect categories found in electrocatalysts and the strategies employed for their introduction or modification, encompassing both direct synthesis and post-treatment methods. We critically assess the merits and limitations of these strategies.

Subsequently, we delve into the realm of microscopy and spectroscopy techniques, shedding light on how they enable us to investigate defects and their roles within electrocatalysts. We also highlight the significance of computational modeling and simulations in gaining a profound understanding of defective behavior. Finally, we address the existing challenges and provide insights into the prospects of engineering, characterizing, and comprehending defects within electrocatalysts for water splitting.

2. Types of Defects

There are various types of defects, including point defects (vacancies, interstitials), line defects (dislocations), planar defects (grain boundaries), and bulk defects. Each type has distinct effects on the catalyst's properties.

2.1 Point Defects

Point defects are atomic-scale irregularities or deviations from the ideal crystal lattice structure in a solid material. These defects occur at specific points or sites within the crystal lattice and can have a significant impact on the material's properties. There are several types of point defects, including (Figure 1):

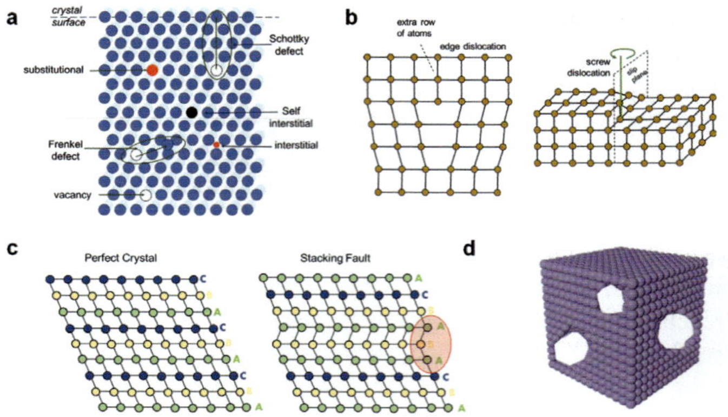

Figure 1. Schematic illustration of various defects including point, line, planar, and bulk defects.

Vacancies: Vacancies are missing atoms at lattice sites. They result from the absence of an atom where one should be. Vacancies can create localized regions with altered electronic properties and affect the material's electrical and thermal conductivity.

Interstitials: Interstitials are extra atoms that occupy positions in the crystal lattice that are not part of the regular lattice structure. These extra atoms can introduce lattice distortions and can also act as charge carriers in semiconductors.

Substitutional Defects: Substitutional defects occur when an atom is replaced by a different type of atom. This substitution can lead to changes in the material's properties, such as electrical conductivity or mechanical strength.

Frenkel Defects: Frenkel defects involve the migration of an atom from its regular lattice position to an interstitial site. This type of defect is common in ionic solids and can result in increased electrical conductivity [35].

Schottky Defects: Schottky defects are specific to ionic compounds and occur when equal numbers of cations and anions are missing from their regular lattice sites. This leads to the formation of vacancies in the crystal structure and can affect the material's electrical and ionic conductivity [36].

2.2 Line Defects

Line defects, also known as one-dimensional defects or dislocations, are structural irregularities that occur in crystalline materials along lines or one-dimensional regions. These defects have a significant impact on the mechanical and electrical properties of materials. There are two primary types of line defects:

Edge Dislocations: Edge dislocations occur when an extra half-plane of atoms is introduced into the crystal lattice. This extra half-plane creates a one-dimensional line defect along the edge of the material. Edge dislocations can be envisioned as a boundary between regions of different atomic packing, leading to a localized lattice distortion. Edge dislocations affect the material's mechanical properties and can act as barriers for the motion of crystal dislocations [36,37].

Screw Dislocations: Screw dislocations result from the shear or twisting of atomic planes within the crystal lattice. These dislocations form a helical or spiral-like structure along the dislocation line, resulting in a one-dimensional defect that affects the material's mechanical behavior [32].

2.3 Planar Defects

Planar defects, also known as two-dimensional defects, are irregularities that occur within the crystal lattice of materials along specific planes or two-dimensional regions. There are several types of planar defects; including grain boundaries – grain boundaries are interfaces that separate different crystal grains or domains within a polycrystalline material. Atomic boundaries, the atomic arrangement changes abruptly from one grain to another, twin Boundaries: twin boundaries occur when a crystal lattice is perfectly mirrored or twinned across a plane. This twinning can lead to a crystal having two regions with opposite orientations, stacking faults: stacking faults are regions within a crystal lattice where there is a deviation from the regular stacking sequence of atomic planes, antiphase boundaries: antiphase boundaries occur in materials with ordered crystal structures, such as intermetallic compounds. At these boundaries, there is a phase shift in the atomic arrangement.

2.4 Bulk Defects

Bulk defects, also known as three-dimensional defects, are irregularities and imperfections that occur within the interior or bulk of a crystalline material rather than at its surfaces or along specific planes. These include pores, which come in a range of sizes (micro, meso, and macropores), as well as cracks, representing localized structural gaps. Inclusions, involving the incorporation of foreign particles into the host crystal, and the presence of distinct phases are also among the bulk defects.

3. Water Splitting via Defect Engineered Electrocatalysts for the HER and OER

3.1 Type of Defects in the Electrodes for OER and HER

In the context of electrocatalysts for the HER and the OER, several types of defects have been identified as factors that boost their catalytic performance.

These include point defects, which can involve the absence or substitution of specific atoms, line defects, which are irregularities along one-dimensional regions, planar defects, which are structural anomalies across two-dimensional planes and bulk defects.

Anion vacancies refer to the absence of anions (negatively charged ions) at specific lattice sites within a material. They are indeed a type of point defect in the crystal lattice of a material, where an anion is missing from its regular position. Point defects, including anion vacancies, can significantly influence the electronic, structural, and catalytic properties of materials. In the context of the HER and OER, anion vacancies play a vital role. These vacancies can alter the electronic structure and surface properties of the electrocatalyst, influencing its catalytic activity. The presence of anion vacancies like oxygen, sulfur, phosphorus etc. can introduce active sites for adsorption and reaction, thus enhancing the kinetics of the HER and OER processes.

The prevalence of oxygen vacancies stands out prominently among anion vacancies, and researchers have actively induced oxygen vacancies through defect engineering in electrocatalysts employed in water splitting reactions. This phenomenon is particularly observed in materials like transition metal oxides for the HER and rutiles, perovskites, layered double hydroxides (LDHs), and spinels for the OER [77–79]. Various techniques, including thermal treatment, etching, and doping, have proven effective in generating oxygen vacancies. These vacancies play a pivotal role in adjusting both geometric and electronic structures, thereby optimizing crucial parameters influencing catalytic performance. It is noteworthy, however, that the presence of oxygen vacancies may also yield adverse effects on material properties, such as an increase in the band gap.

Beyond the investigation of oxygen vacancies, significant attention is directed towards sulfur vacancies due to their crucial role in transition metal chalcogenides, particularly in the context of the HER [80]. In materials containing sulfur, the absence of sulfur ions leads to the formation of sulfur vacancies. These vacancies have become a focal point in catalytic research, especially within sulfur-containing compounds. Defect engineering strategies aimed at introducing sulfur vacancies into materials like molybdenum disulfides and other sulfides have proven effective in enhancing HER activity. Notably, the activation and optimization of MoS_2, a promising HER catalyst with inert basal planes, has been achieved through the deliberate introduction of sulfur vacancies [81,82].

Like anion vacancies, researchers have extensively engineered cation defects to boost the catalytic performance of catalysts in both the HER and the OER [83–85]. Cation defects involve the absence of cations (positively charged ions) at specific sites within a crystal lattice. These defects can influence the electronic and catalytic properties of materials. Some examples of cation vacancies include (i) Transition Metal Vacancies: In transition metal-based electrocatalysts, the absence of specific transition metal ions from their regular lattice positions creates transition metal vacancies. These vacancies can significantly impact the catalytic activity of the material. (ii) Metal Oxide Vacancies: In metal oxide electrocatalysts, such as perovskite oxides, the absence of metal cations (e.g., La, Sr, Mn) results in metal oxide vacancies. These vacancies can introduce active sites for electrochemical reactions. Creating cation defects in transition metal-based electrocatalysts can be achieved through various strategies like thermal treatment, ion implantation, chemical etching, doping.

Beyond point defects, which encompass vacancies, other types of defects such as line defects (e.g., edge dislocation and screw dislocation), planar defects (e.g., twin boundaries, grain boundaries, and stacking faults), and bulk defects (e.g., pores) can also positively influence the catalytic performance of electrocatalysts utilized in water splitting. For example, bulk defects can lead to heightened porosity, thereby exposing a greater number of active sites [86–90].

3.2 Defects Engineering Strategies

To comprehend the significance of defects, we initially outline defect engineering strategies aimed at elucidating the processes through which defects are incorporated into catalysts for the HER and OER. Defect engineering can be broadly categorized into two groups: the direct synthesis of defective catalysts and the post-treatment of catalysts to induce defects. Detailed strategies, along with their respective strengths and weaknesses, are presented in Figure 3. Certain approaches exhibit versatility in both categories, serving as means for introducing defects through direct synthesis and post-treatment. Examples include doping/substitution, hydrothermal/solvothermal methods, and ball milling. The subsequent classification is established on the prevailing circumstances.

Through effective defect engineering, catalysts exhibit a substantial improvement in catalytic performance, as detailed in Tables 1 and 2. The activities of various catalysts, with and without defects, are further summarized concerning overpotential at 10 mA cm^{-2}, Tafel slope, and catalyst mass loading, presented in Table 2. Notably, as evidenced in Table 2, the presence of vacancy defects appears to more effectively enhance catalytic activity, particularly for OER catalysts featuring oxygen vacancies.

3.2.1 Direct Synthesis of Defective Catalysts

3.2.1.1 Doping/Substitution

The utilization of doping/substitution represents another widely adopted approach in defect engineering, leading to the creation of substitutional defects and interstitial defects, as illustrated in Figure 2a. This strategy involves introducing heteroatom precursors or employing alternative treatments, such as heat treatment in a specific atmosphere like NH_3. The heteroatoms introduced can be either isovalent or aliovalent cations/anions. Notably, aliovalent cation doping/substitution can induce the formation of anion vacancies and alter the valence of cations in other sites. For example, in the case of perovskite, the partial replacement of A-site cations with aliovalent cations may generate oxygen vacancies and modify the valence of B-site cations to maintain charge neutrality.

Table 1. A summary of defective catalysts covering categories of defects, approaches for structural modification, characterization techniques used for investigating defects, and their electrochemical applications

Catalyst	Defect type	Synthesis technique	Characterization technique	Reaction	Reference
Fe-Ni_2P (P_v)	P vacancies	$NaBH_4$ reduction	XPS	OER	[38]
cd-$Ni(OH)_2$–$NaBH_4$	Cationic and anionic defects	Alkaline corrosion, $NaBH_4$ reduction	XPS, XANES, EXAFS	OER	[39]
Mn_6-CoO	Oxygen vacancies	Mn doping	XPS, ESR, XANES	HER, OER, water electrolysis	[40]
$NiCo_2O_4$	Oxygen defects	lithiation	XPS, EPR spectroscopy	OER	[41]
DE-TDAP	Structural defects (micro/mesoporous structure)	N, S, O tri-doped	XPS	OER, HER	[42]
FeOOH	Surface defects	Hydrothermal	XPS, DFT	OER	[43]
Fe-N-CHPS	Plane defects	Ultra-low metal doping	RAMAN, XPS	OER, HER	[44]
Cobalt based spinel oxides	Oxygen vacancies	$NaBH_4$ reduction	PLS, XPS	OER	[45]
Co_3O_4	Oxygen vacancies	$NaBH_4$ reduction	XPS	OER	[46]
Sr_2CoFeO_6	Surface defects	Acid treatment	XPS, ICP, EDS	OER	[47]
Co_3S_4@Co_3O_4/NSC	Mesoporous, structural defects	Calcination	RAMAN, XPS	OER	[48]
RuO_2	Cation vacancy	Acid etching	XPS, XANES, EXAFS	OER	[49]
$CoZnRuO_x$	Oxygen vacancies	Electrochemical etching	XPS	OER, water-splitting	[50]

Contd...

Contd...

Catalyst	Defect type	Synthesis technique	Characterization technique	Reaction	Reference
RuO_2	Strain engineering	Solid source Metal-organic molecular beam epitaxy	NA	OER	[51]
O_v-Fe MOF/IF	Anion defects	Solvothermal and plasma engraving	XPS, XANES, EXAFS, EPR	HER, OER, water electrolysis	[52]
NC@Vo·-RuO_2/CNTs-350	Oxygen vacancy	Hydrothermal and calcination	XPS, EPR,	OER, water electrolysis	[53]
NiFeZn MOF	Zn defects	solvothermal	XPS	OER	[54]
Al_2Cu-MOF@M-OOHv-ER/CF	Oxygen defects	Electrochemical etching	RAMAN	OER	[55]
NiFe-BTC/IL-PF_6			EPR, XPS	OER	[56]
$Co_3O_4/CoMoO_4$	Oxygen vacancy	Hydrothermal	XPS, RAMAN	OER	[57]
Co_3O_4-V_{Co}-OER	Co defects	Thermal oxidation	DFT, RAMAN, XRD	OER	[58]
CoP/Co_2P	P vacancies	Molten-salt-mediated phosphidation	EPR, RAMAN, XPS	HER, OER, water electrolysis	[59]
Defect rich-FeOOH	Oxygen vacancies	Solvothermal and Electron beam radiation	XRD, RAMAN, XANES	OER	[60]
V-$CuCo_2O_4$/CF	Oxygen vacancies, Co^{2+} defects	Laser ablation	EPR, RAMAN, XANES	OER	[61]
Co, Mo-NiFe LDH	Cation vacancy	Hydrothermal (Co dopant)	XPS, EXAFS	OER	[62]
WS_2/MoS_2	Mo vacancy	Hydrothermal	DFT	OER	[63]

V-Ni MOF@ FeOOH	Oxygen vacancy	Hydrothermal, alkali etching treatment	XPS	OER	[64]
Co_3S_4	Sulfur vacancy	Annealing	EPR, XPS	OER	[65]
Co-Ni_2P/NF-corr	NA	H_2SO_4 etching, phosphorization	XPS	OER	[66]
Ni,Co,Yb-FeOOH	Oxygen vacancy, edge dislocation	Hydrothermal	XRD, XPS, EPR	OER	[67]
$(FeCoNiB_{0.75})_{97}Pt_3$	Lattice defects and stalking fault defects	Electrochemical dealloying	XANES, EXAFS	OER, HER	[68]
d-(Fe,Ni)OOH	Oxygen vacancy, multidimensional defects, atomic defects	Room temperature stirring	RAMAN, EPR, XPS	OER, water splitting	[69]
TaS_2	Sulfur vacancies	Electrochemical intercalation, Desulfurization	XPS, EPR,	HER, OER	[70]
$Bi_{0.15}Sr_{0.85}Co_{1-x}Fe_xO_{3-6}$	Oxygen vacancies	Ball-milling	XPS	OER	[71]
V,P-CoS_2/CC	Co defects	Annealing	EPR, XPS	OER, overall water splitting	[72]
Mo-Co_9Se_8@NiSe/ NF-60	Se vacancies	Hydrothermal	EPR, XPS	OER, overall water splitting	[73]
P doped CoS	S vacancies	P dopant (Annealing in tube furnace)	XPS, XRD	OER	[74]
Ru/Rh-doped FeOOH	Oxygen vacancies	One-step oil bath method	EPR, XPS	OER	[75]
Co/Co_9S_8/MnS-NMC	NA	S-dopant hydrothermal	XPS, RAMAN	OER	[76]

Table 2. Comparing the catalytic performance of different catalysts in the presence and absence of defects for the HER and OER

Catalyst	Electrolyte	Electrode	η (mV) with and without defects	Tafel slope (mV/dec) with and without defects	Mass loading (mg/cm^2)	application	Reference
Fe-Ni_2P (P_v)	1 M KOH	GCE	290 *vs* 342 at 10 mA cm^{-2}	41.15 vs 94.88	NA		[38]
cd-$Ni(OH)_2$–$NaBH_4$	1 M KOH	Carbon paper	205 vs 270 at 10 mA cm^{-2}	31.1 vs 47.3	4		[39]
Mn_6-CoO	1 M KOH	Ni foam	301 vs 427 at 10 mA cm^{-2}	96.7 vs 172.5	NA	OER	[40]
			25.6 vs 177.6 mV	72.6 vs 119.1		HER	
$NiCo_2O_4$	1 M KOH	Carbon cloth	270 vs 344 at 50 mA cm^{-2}	46 vs 81	NA	OER	[41]
DE-TDAP	0.1 M KOH	GCE	346 vs 498 154 vs 396	67 vs 169 83 vs 148	NA	OER HER	[42]
FeOOH	1 M KOH	Carbon cloth	342 vs 545	77.8 vs 240.3	2	OER	[43]
Fe-N-CHPS	1 M KOH	GCE	248 vs 270 172 vs 259	48 vs 52 30 vs 77	0.2	OER HER	[44]
Cobalt based spinel oxides	1 M KOH	GCE	376 vs 378	51 vs 53	NA	OER	[45]
Co_3O_4	0.1 M KOH	GCE	378 vs 477	58.18 vs 79.73	20	OER	[46]
Sr_2CoFeO_6	1 M KOH	GCE	300 vs 396	59.62 vs 119.5	0.397	OER	[47]

Co_3S_4@Co_3O_4/NSC	1 M KOH	GCE	312 vs 347	62.8 vs 162	0.340	OER	[48]
$CoZnRuO_x$	1 M KOH	GCE	244 vs 373	81.6 vs 86.7	6	OER	[50]
O_v-Fe MOF/IF	1 M KOH	Fe foam	200 vs 343 72 vs 182	44 vs 54 65 vs 151	NA	OER HER	[52]
NC@Vo·-RuO_2/CNTs-350	0.5 M H_2SO_4	GCE	170 vs 308	38.9 vs 69.7	0.25	OER	[53]
NiFeZn MOF	1 M KOH	GCE	233 vs 284	37.8 vs 78.7	NA	OER	[54]
Al_2Cu-MOF@M-OOHv-ER/CF	1 M KOH	Cu foam	366 vs 428	22.3 vs 45.5	NA	OER	[55]
NiFe-BTC/IL-PF_6	1 M KOH	Ni sheet	210 vs 370	103 vs 258	NA	OER	[56]
Co_3O_4/$CoMoO_4$	1 M KOH	GCE	342 vs 580	72 vs 143	0.71	OER	[57]
Co_3O_4-V_{Co}-OER	1 M KOH	GCE	268 vs 361	60.5 vs 117.7	NA	OER	[58]
CoP/Co_2P	1 M KOH	Ti mesh	300 vs 351 190 vs 300	68 vs 91 84 vs 116	2	OER HER	[59]
Defect rich-FeOOH	1 M KOH	Carbon paper	307 vs 372	36 vs 63	0.192	OER	[60]
V-$CuCo_2O_4$/CF	1 M KOH	Carbon fiber	204 vs 353	40.7 vs 65.3	2.73	OER	[61]
V-Ni MOF@FeOOH	1 M KOH	Ni foam	370 vs 449	79 vs 148	NA	OER	[64]
Co_3S_4	1 M KOH	GCE	270 vs 380	59 vs 85	NA	OER	[65]
Co-Ni_2P/NF-corr	1 M KOH	Ni foam	267 vs 296	64 vs 125	NA	OER	[66]

Contd...

Contd...

Catalyst	Electrolyte	Electrode	η (mV) with and without defects	Tafel slope (mV/dec) with and without defects	Mass loading (mg/cm^2)	application	Reference
Ni,Co,Yb-FeOOH	1 M KOH	Carbon cloth	230.9 vs 334.6	22.7 vs 41	6.6	OER	[67]
$(FeCoNiB_{0.75})_{97}Pt_3$	1 M KOH	As-spun ribbon	170 vs 338	37.8 vs 53	NA	OER	[68]
d-(Fe,Ni)OOH	1 M KOH	Fe foam	213 vs 240	21.7 vs 36.7	NA	OER	[69]
TaS_2	1 M KOH 0.5 M H_2SO_4	Ni foam	153 vs 270 99 vs 176	38 vs 78 39 vs 48	NA	OER HER	[70]
$Bi_{0.15}Sr_{0.85}Co_{1-x}Fe_xO_{3-6}$	0.1 M KOH	GCE	330 vs 436	61.9 vs 110.7	0.255	OER	[71]
V,P-CoS_2/CC	1 M KOH	Carbon cloth	227 vs 362	66 vs 77	5.05	OER	[72]
Mo-Co_9Se_8@NiSe/NF-60	1 M KOH	Ni foam	343 vs 399	61.86 vs 93.88	NA	OER	[73]
P doped CoS	1 M KOH	GCE	280 vs 332	79.38 vs 84.25	NA	OER	[74]
Ru/Rh-doped FeOOH	1 M KOH	Carbon paper	223/230 vs 356	63.6/67.9 vs 121.8	1	OER	[75]
Co/Co_9S_8/MnS-NMC	1 M KOH	N-doped mesoporous carbon substrate	330 vs 365	55.1 vs 75.7	1	OER	[76]

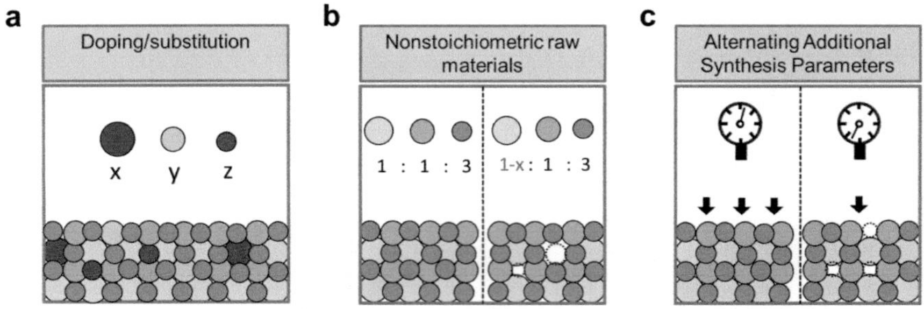

Figure 2. Schematic illustration of the different strategies of direct synthesis of defective catalysts, (a) doping or substitution, (b) nonstoichiometric raw materials, (c) alternate additional synthesis parameters.

Doping/substitution with cations (such as precious metal, alkali/alkaline metal, and transition metal) or anions (such as sulfur and halogen) has been extensively studied in various electrocatalysts, including perovskites, spinels, layered structures, and rock salts. The impact on the inherent physicochemical properties is contingent upon the type of dopant introduced. Dopants can effectively tune the geometric and electronic structures of catalysts, directly influencing their activity and stability, as elaborated in Tables 1 and 2.

For example, (FeNi)O/NiSe@NF with oxygen vacancies was synthesized via Fe-incorporation to enhance the water splitting activity by an electrochemical method [91]. The formation of oxygen vacancies was ascribed to the Fe incorporation. There are findings indicating that the OER activities of NiO nanocrystals improve significantly when doped with Li^+. This enhancement is attributed to the adjusted Fermi levels, increased hybridization strength between Ni 3d and O 2p, and a reduced energy band gap compared to the original, undoped NiO [92].

A perovskite, fluorine-doped lanthanum cobaltate (F_x-$LaCoO_3$), was synthesized using traditional sol-gel method for OER [93]. The subsequent characterization results showed that heteroatom doping enhances intrinsic catalytic activity to induce lattice distortion and optimize spin-dependent orbital interaction to alter charge transfer between catalysts and reactants.

3.2.1.2 Starting with Non-stoichiometric Raw Materials

Commencing with non-stoichiometric raw materials proves to be an efficient method for directly instigating defects in electrocatalysts, encompassing materials like perovskite oxides, metal phosphides, and others. Creating defects through non-stoichiometric raw materials involves intentionally introducing a deviation from the ideal chemical composition, leading to the formation of vacancies or excesses of certain elements within the material (Figure 2b). This approach is commonly used in defect engineering to tailor the properties of materials, including catalysts. Here's how the process generally works:

(i) Identify Raw Materials: Begin with non-stoichiometric raw materials, which means the starting materials have a different ratio of elements than the ideal or stoichiometric composition. (ii) Introduce Vacancies: If the goal is to create vacancies (missing atoms or ions) within the material, intentionally leave out a certain percentage of one type of atom during the synthesis process. This can result in spaces or "holes" in the crystal lattice structure. (iii) Excess of Certain Elements: Alternatively, one can intentionally introduce an excess of certain elements. This excess can disrupt the regular arrangement of atoms in the crystal lattice, leading to the creation of defects. (iv) Controlled Synthesis Conditions: The key is to carefully control the synthesis conditions, such as temperature, pressure, and reaction time, to ensure that the non-stoichiometric composition is maintained during the material's formation.

Specifically, during the synthesis of catalysts such as perovskite oxides, characterized by chemical formulas of ABO_3, the cation deficiency can be adjusted by employing non-stoichiometric raw materials with molar ratios of A/B = 1:(1 ± x), as illustrated in Figure 3a. In this scenario, anion vacancies, such as oxygen vacancies in ABO_3, frequently emerge as integral components of the lattice to maintain charge neutrality. The efficacy of this approach lies in its simplicity. Furthermore, defects formed in this manner exhibit a more uniform distribution in both the surface layer and the bulk, in contrast to those obtained through post-treatment methods. Careful tuning of precursor proportions is essential to prevent the generation of a second phase. However, this strategy has limitations in its applicability, as it is not suitable for materials like metal LDHs.

Non-stoichiometric NiO nanosheets containing plentiful Ni vacancies (Ni_xO) formed on the surface of NiSe nanorods (Ni_xO/NiSe@NF) through an *in situ* electrochemical oxidation process. The introduction of a nickel vacancy in the crystal structure of NiO has the effect of elevating the free energy of *OH, consequently lowering the theoretical overpotential of the Ni_v-NiO electrocatalyst [91].

3.2.1.3 Altering Additional Synthesis Parameters

Defects can be induced by manipulating various synthesis conditions, including pressure, temperature, and atmosphere, as depicted in Figure 2c. This approach allows for the deliberate engineering of defects; however, its applicability is constrained. For example, adjustments in synthesis pressure typically led to the formation of anion vacancies.

3.2.2 Post-treatment of Catalysts to Induce the Formation of Defects

3.2.2.1 Heat Treatment

Due to its straightforward applicability, heat treatment (illustrated in Figure 3a) has found widespread use in the treatment of various materials, including metal oxides, metal dichalcogenides, metal phosphides, and others, for the purpose of inducing vacancies and bulk defects. During thermal treatment, the experimental parameters that can be adjusted include temperature, duration, and

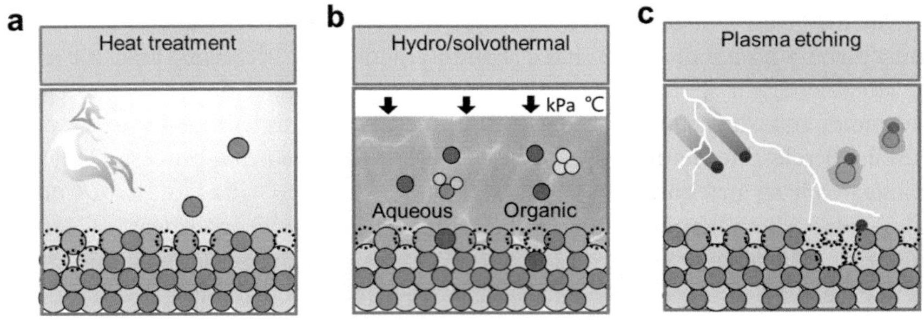

Figure 3. Schematic illustration of the different post-treatment strategies to form defective catalysts: (a) heat treatment, (b) hydro/solvothermal, (c) plasma etching.

the surrounding atmosphere. These variations in experimental conditions play a significant role in determining whether defects are formed, their type, quantity, and the rate at which they are generated. For example, applying heat treatment to eliminate sulfur atoms from the surface layer of molybdenum disulfide resulted in the formation of sulfur vacancies at temperatures below 600 °C. Beyond 600 °C, a substantial removal of sulfur atoms ensued.

3.2.2.2 Hydrothermal/Solvothermal Treatment

Hydrothermal and solvothermal methods can be employed both as direct synthesis techniques for defective catalysts and as post-treatment strategies to induce defects in materials that have been already synthesized. In the direct synthesis approach, hydrothermal or solvothermal methods involve the reaction of precursors in a high-temperature and high-pressure aqueous or solvent environment. Defects are introduced during the formation of the crystal lattice in these harsh conditions (Figure 3b). For instance, non-stoichiometric ratios of reactants or the presence of impurities in the reaction mixture can lead to the incorporation of defects in the final product. Hydrothermal synthesis of metal oxides, such as Fe doped Co_3O_4, often results in the formation of oxygen vacancies and other defects in the crystal structure.

After the initial synthesis, the material is subjected to a hydrothermal or solvothermal post-treatment. This involves exposing the material to elevated temperatures and pressures in a controlled environment. The post-treatment can create defects by altering the existing crystal lattice through processes like re-crystallization, surface modification, or the removal/addition of certain elements. Introduction of foreign elements during hydrothermal post-treatment can lead to defects, as atoms of different sizes may replace the original atoms in the crystal lattice, causing distortions and vacancies.

3.2.2.3 Plasma and Chemical Etching

Plasma etching and chemical etching are two techniques used for creating defects in electrode materials. These methods are employed to selectively remove material or modify its surface, leading to the introduction of defects in the material structure. Plasma etching involves the use of plasma, which is a partially ionized gas, to remove material from a substrate through chemical reactions. During plasma etching, the material to be etched is exposed to a low-pressure plasma. Reactive ions in the plasma chemically react with the material's surface, leading to the removal of atoms or molecules (Figure 3c). Examples of plasma etching gases are Ar, N_2, and NH_3 etc. Ar-plasma (Ar-P) etching strategy was employed to fabricate a hierarchical structure of Mo-substituted and oxygen vacancy-enriched spinel Co_3O_4 porous nanoneedle arrays. These were grown *in situ* on carbon cloth, denoted as P-Mo-Co_3O_4@CC, with the aim of enhancing the OER performance. The Ar-plasma technology proved effective in creating vacancy sites on the hydroxide surface, promoting the electrostatically driven anchoring of Mo anion salts. This, in turn, facilitated the substitution of Mo atoms and the generation of oxygen vacancies on the surface of Co_3O_4 [94].

Chemical etching involves the use of chemical solutions to selectively remove material from the surface of a substrate. The material is immersed in an etchant, which reacts with the surface and dissolves or loosens the material, creating defects. Examples of chemical etchants are HF, HCl, KOH, NaOH, and HNO_3, etc.

4. Characterization of Defects

Advanced analytical methodologies, specifically employing spectroscopy and microscopy techniques, are used for analyzing the engineered defects in the electrocatalysts. These methodologies are vital for unraveling the significance of defects and elucidating the origin of performance differences observed in electrochemical reactions The effect of defects in electrocatalysts on the electronic distribution and the atomic structure should not be ignored. The different characterization techniques employed for analyzing defects are described below.

4.1 Raman Spectroscopy

Raman spectroscopy is a surface-sensitive technique, which offers examining a range of defects including vacancies and dopants. Raman Spectroscopy has been widely used to analyze and characterize chemical bonding energy on catalyst surfaces. Defects present in materials may change vibrational modes and Raman peaks may shift or newly appear [95]. Confirmation of the formation of cation vacancy defects and the movement of the surface through the metal oxide interface during the OER has also been established using Raman spectroscopy. Wang et al. doped Ru (Ir) into self-supported NiV LDH to accelerate water splitting kinetics [96]. From Raman spectra, a certain degree of redshift and lower intensities were observed in Ru-doped NiV LDH (Figure 4a). This shift is attributed to the size effect or doping effect and defects. *In situ* Raman spectroscopy can provide more accurate information regarding the formation of defects. Louie et al. carried out detailed investigations using *in situ* Raman spectroscopy of electrodeposited Ni-Fe films for OER [97]. Raman spectra

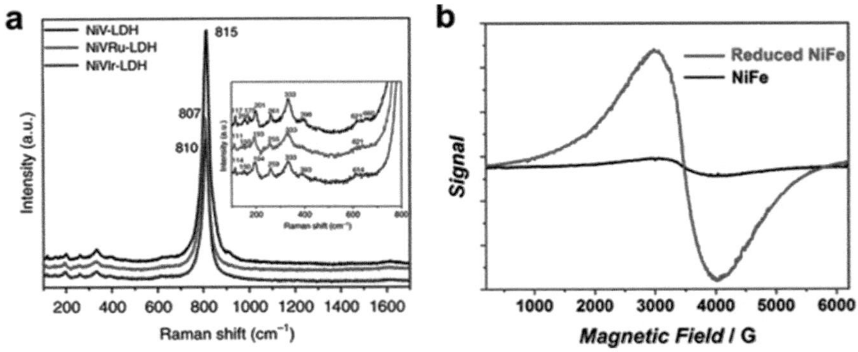

Figure 4. Characterization of defects (a) Raman spectra of NiV-LDH, NiVRu-LDH, and NiVIr-LDH with 532 nm excitation under ambient air. (Copyright 2020, Springer Nature) (b) EPR spectra. (Copyright 2018, American Chemical Society)

indicate that, under OER potentials, Ni in the Ni–Fe possesses a structural unit similar to NiOOH. Moreover, the presence of Fe significantly alters the local environment of Ni–O, as shown by the relative intensities of the two NiOOH bands at 475 and 555 cm^{-1}.

4.2 Electron Paramagnetic Resonance (EPR)

EPR spectroscopy is generally used to measure the unpaired electrons in catalyst materials, due to coordination unsaturated structures such as vacancy defects. Recently, many studies have used EPR spectroscopy to confirm the presence of oxygen vacancies and sulfur vacancies in the catalyst materials. The intensity of the EPR signal could be affected by the number of oxygen vacancies present in the materials. Asnavandi et al. prepared NiFe with abundant oxygen vacancies by a facile and economic $NaBH_4$ reduction process [98]. The EPR spectroscopy confirms the presence of oxygen vacancies in NiFe after $NaBH_4$ treatment with a strong increase in EPR signal at g = 1.99 as shown in Figure 4b.

4.3 X-ray Photoelectron Spectroscopy (XPS)

XPS is extensively employed for surface analysis of materials, and the capability to detect defects, including vacancies and dopants. By analyzing binding energy (BE), XPS can detect defects by variations in elemental compositions, chemical states, and chemical environments. Ma et al. created the oxygen vacancies in Co_3O_4 using the Ar plasma treatment and used XPS to prove it [99]. Xie and coworkers prepared spinel-structure $NiCoO_4$ with oxygen vacancies for the water oxidation reaction [100]. O 1s spectra, the peak at 531.2 eV assigned to oxygen vacancies, while the peaks at 529.7 and 532.6 eV are associated with oxygen-metal bond and hydroxyl species of surface-absorbed water molecules respectively.

4.4 X-ray Absorption Spectroscopy (XAS)

XAS serves as a highly valuable tool for studying the electronic structures and local geometries of atoms in materials. The XAS spectrum consists of two distinct regions: X-ray Absorption Near-Edge Structure (XANES) and Extended X-ray Absorption Fine Structure (EXAFS). XANES demonstrates high sensitivity to valence states and symmetry, whereas EXAFS allows for the exploration of short-range order, focusing on factors such as coordination number and local atomic distance. This information is crucial for supporting the analysis of defects in the catalysts. Cai et al. compared the pre-edge energy of Co K-edge XANES of pristine Co_3O_4 and reduced Co_3O_4, a shift to lower energy was observed in reduced Co_3O_4 compared to pristine Co_3O_4, suggesting a lower oxidation state of Co and existence of oxygen vacancies [101]. Wang and coworkers also used the XANES spectrum to systematically investigate defective Co_3O_4 samples [102].

EXAFS provides more detailed information about defects compared to XANES. Liu and coworkers used Fourier-transformed EXAFS spectra to confirm the existence of oxygen vacancies in CoFe-LDH [103]. Recently, *in situ/operando* XAS has proven effective in offering more detailed insights to elucidate the role of defects during electrochemical reactions. Choi and coworkers used *in situ/operando* XAS to investigate the cation vacancies present in NiFe-O nanoparticles [104].

4.5 Transmission Electron Microscopy (TEM)

The two basic imaging modes of TEM, bright field (transmitted beam) and dark field (diffracted beam), can observe the various defects in the electrocatalysts. Additionally, both TEM and scanning TEM (STEM) can observe defective structures at lower resolution. Moreover, they can obtain local atomic structure information, including interplanar spacing and atomic arrangement, at a lower resolution, while at a higher resolution, they can discern local defects such as edge dislocation, grain boundary, vacancy, and lattice disorder.

The various defects including mesopores in the nitrogen-doped carbon nanosheets were detected by TEM analysis [105]. Additionally, edge dislocation in Co_3O_4 [106], cation vacancies in NiCu LDH [107], stacking faults, and twins in Ni_3FeN on N-doped graphene [0.1002/adfm.201706018.] as well as oxygen vacancies in $SrCoO_2$ [10.1038/ncomms11053] were also observed using TEM techniques. In addition, to the above-mentioned characterization techniques, several other techniques including AFM, FTIR, and SEM have been employed to explore the defects.

5. Role of Defects

Defects play a crucial role in the field of electrochemical water splitting, encompassing both the Oxygen Evolution Reaction (OER) and the Hydrogen Evolution Reaction (HER). These defects can significantly influence the catalytic activity, kinetics, and overall performance of electrocatalysts.

5.1 Enhanced Catalytic Activity

Defects, such as vacancies and lattice distortions, can create active sites on the catalyst surface, promoting the adsorption and reaction of oxygen species during the OER. These active sites reduce the overpotential required for the OER, enhancing overall catalytic activity. Defects can facilitate the adsorption of hydrogen intermediates, reducing the energy barrier for hydrogen evolution during the HER. This results in improved catalytic efficiency and lower overpotential requirements.

5.2 Improved Reaction Kinetics

Defects can alter the electronic structure and surface energy of the catalyst, influencing the kinetics of oxygen evolution. This modification enhances the rate of reaction and contributes to faster kinetics during OER (Figure 5a). Defects can provide suitable binding sites for hydrogen adsorption and desorption, thereby influencing the reaction kinetics of hydrogen evolution. This is particularly important for achieving high reaction rates.

5.3 Optimized Adsorption Energy

Defects can modulate the adsorption energy of reaction intermediates, ensuring the appropriate binding strength for OER and HER. This optimization is crucial for achieving an efficient and balanced water splitting process.

5.4 Facilitated Charge Transfer

The presence of defects, such as oxygen vacancies, can enhance the charge transfer kinetics at the catalyst-electrolyte interface. This facilitates the transfer of electrons during OER and HER, leading to improved electrocatalytic performance (Figure 5b, c).

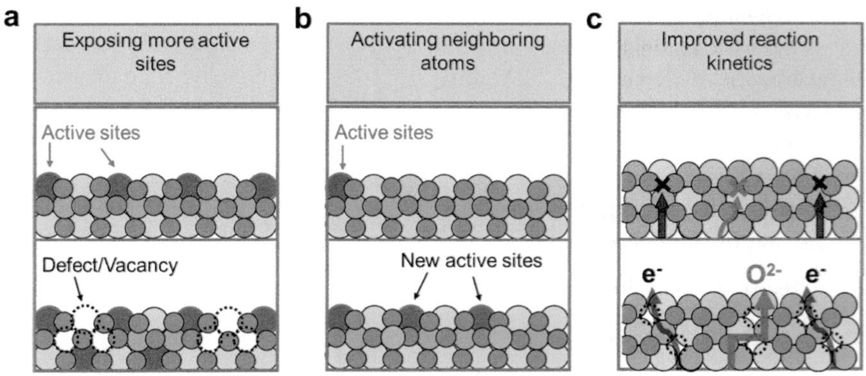

Figure 5. Schematic illustration of the role of defects in electrocatalysts toward the HER and OER: (a) exposing more active sites, (b) activating neighboring atoms, (c) improved reaction kinetics.

5.5 Increased Surface Reactivity

Defects create active sites with higher surface reactivity, allowing for increased interactions with water molecules and reaction intermediates. This heightened reactivity contributes to the overall efficiency of electrochemical water splitting.

5.6 Stability and Durability

Certain types of defects can contribute to the stability and durability of electrocatalysts by preventing agglomeration or detachment of active sites. Stable defective structures can lead to long-term catalytic performance.

5.7 Tailored Selectivity

Defect engineering enables the tailoring of electrocatalysts for specific reactions, influencing the selectivity of the OER and HER. This is essential for designing catalysts that selectively promote either oxygen or hydrogen evolution.

5.8 Synergistic Effects

The synergy between different types of defects, such as metal vacancies and heteroatom doping, can create unique active sites with enhanced catalytic properties. This synergistic effect is advantageous for optimizing electrocatalytic performance.

6. Challenges and Perspectives

In conclusion, we have studied recent advancements in the field of defect engineering applied to electrocatalysts for water splitting. We provided an overview of various defect types and the strategies employed in defect engineering. Following this, we discussed microscopy and spectroscopy characterization techniques utilized for the analysis of defects to uncover their roles. We systematically explored the crucial functions of defects, including the augmentation of active sites, electronic structure modification, tailored selectivity, and synergistic effect of metal vacancy and heteroatom doping. Despite notable achievements in defect engineering, there are still significant challenges that need to be addressed, including predicting performance, refining precise defect engineering, conducting mechanism studies, addressing stability concerns, and employing *in situ/operando* characterizations.

1. Precisely controlling the synthesis of defective electrocatalysts remains a challenge. Achieving a uniform distribution of defects and tailoring their types and densities are crucial for optimizing catalytic performance. Accurately characterizing and quantifying defects in electrocatalysts pose challenges. High-resolution techniques are often required, and interpreting results can be complex due to the presence of various defect types.

2. Some defects may lead to increased instability and reduced durability of electrocatalysts over time. Balancing defect-induced activity with long-term stability remains a challenge in the design of practical catalysts. Establishing clear relationships between specific types of defects and their influence on catalytic activity is a complex task. Defects may have different impacts depending on the material and reaction conditions.
3. Tailoring defective electrocatalysts to simultaneously optimize the Oxygen Evolution Reaction (OER) and Hydrogen Evolution Reaction (HER) presents a challenge. Achieving balanced catalytic activity for both reactions is essential for overall water splitting efficiency. Developing scalable and cost-effective methods for synthesizing defect-engineered electrocatalysts is a challenge. Many advanced synthesis techniques may not be suitable for large-scale production.
4. Advances in understanding the relationship between specific defects and catalytic activity will enable more precise and tailored defect engineering. This knowledge will contribute to the rational design of electrocatalysts with enhanced performance. Continued development of advanced multiscale characterization techniques will provide deeper insights into defective structures. This includes combining spectroscopy, microscopy, and computational methods to gain a comprehensive understanding. Integrating theoretical approaches, such as computational modeling and simulations, will aid in predicting the impact of defects on catalytic activity. This will streamline the design process and reduce the need for extensive experimental trials.
5. Combining defect engineering with other strategies, such as heteroatom doping or nano structuring, may offer synergistic effects, leading to enhanced catalytic performance. Hybrid approaches can address multiple challenges simultaneously. Advancements in *in situ* and operando techniques will allow researchers to observe catalyst behavior under working conditions. This real-time monitoring is crucial for understanding the dynamic changes in defective structures during electrochemical water splitting.
6. Developing scalable and economically viable synthesis methods for defect-engineered electrocatalysts is essential for their practical application. This includes exploring simpler techniques suitable for large-scale production. Assessing the environmental and economic impact of defect-engineered electrocatalysts is crucial. A comprehensive understanding of the life cycle and cost-effectiveness will guide the development of sustainable technologies.

References

1. M.S. Dresselhaus, I.L. Thomas, Alternative energy technologies, Nature, 414(6861) (2001) 332-337. https://doi.org/10.1038/35104599.

2. S. Anantharaj, S.R., Ede, K. Karthick, S. Sam Sankar, K. Sangeetha, P.E. Karthik, S. Kundu, Precision and correctness in the evaluation of electrocatalytic water splitting: Revisiting activity parameters with a critical assessment, Energy Environ. Sci., 11(4) (2018) 744-771. https://doi.org/10.1039/C7EE03457A.
3. K. Patil, P. Babar, X. Li, V. Karade, S. Kim, S.Y. Jang, P. Bhoite, J.H. Kim, Co-Fe-B nanochain electrocatalysts for oxygen evolution at high current density, ACS Appl. Nano Mater., 5(5) (2022) 6260-6267. https://doi.org/10.1021/ACSANM.2C00312/ASSET/IMAGES/LARGE/AN2C00312_0006.JPEG.
4. K. Patil, P. Babar, X. Li, J.S. Jang, D. Kim, M. Cheol Baek, P. Bhoite, J. Hyeok Kim, Facile electrodeposited NiMoSe nanospheres for hydrogen evolution reaction, Mater Lett., 310 (2022) 131409. https://doi.org/10.1016/J.MATLET.2021.131409.
5. M. Qu, Y. Jiang, M. Yang, S. Liu, Q. Guo, W. Shen, M. Li, R. He, Regulating electron density of NiFe-P nanosheets electrocatalysts by a trifle of Ru for high-efficient overall water splitting, Appl. Catal. B, 263 (2020) 118324. https://doi.org/10.1016/J.APCATB.2019.118324.
6. P. Babar, K. Patil, J. Mahmood, S.-J. Kim, J.H. Kim, C.T. Yavuz, Low-overpotential overall water splitting by a cooperative interface of cobalt-iron hydroxide and iron oxyhydroxide, Cell Rep. Phys. Sci., 3 (2022) 100762. https://doi.org/10.1016/j.xcrp.2022.100762.
7. K. Patil, P. Babar, D.M. Lee, V. Karade, E. Jo, S. Korade, J.H.Kim, Bifunctional catalytic activity of Ni-Co layered double hydroxide for the electro-oxidation of water and methanol, Sustain Energy Fuels, 4(10) (2020) 5254-5263. https://doi.org/10.1039/D0SE00899K.
8. K. Patil, P. Babar, H. Bae, E. Jo, J.S. Jang, P. Bhoite, S. Kolekar, J.H. Kim, Enhanced electrocatalytic activity of a layered triple hydroxide (LTH) by modulating the electronic structure and active sites for efficient and stable urea electrolysis, Sustain Energy Fuels, 6(2) (2022) 474-483. https://doi.org/10.1039/D1SE01478A.
9. P. Babar, K. Patil, V. Karade, K. Gour, A. Lokhande, S. Pawar, J.H. Kim, In situ fabrication of nickel-iron oxalate catalysts for electrochemical water oxidation at high current densities, ACS Appl. Mater. Interfaces, 13(44) (2021) 52620-52628. https://doi.org/10.1021/ACSAMI.1C14742/ASSET/IMAGES/LARGE/AM1C14742_0007.JPEG.
10. P. Babar, K. Patil, D.M. Lee, V. Karade, K. Gour, S. Pawar, J.H. Kim, Cost-effective and efficient water and urea oxidation catalysis using nickel-iron oxyhydroxide nanosheets synthesized by an ultrafast method, J. Colloid Interface Sci., 584 (2021) 760-769. https://doi.org/10.1016/J.JCIS.2020.09.108.
11. J. Zhu, L. Hu, P. Zhao, L.Y.S. Lee, K.Y. Wong, Recent advances in electrocatalytic hydrogen evolution using nanoparticles, Chem Rev., 120(2) (2020) 851-918. https://doi.org/10.1021/ACS.CHEMREV.9B00248/ASSET/IMAGES/MEDIUM/CR9B00248_0041.GIF.
12. C. Wei, R.R. Rao, J. Peng, B. Huang, I.E.L. Stephens, M. Risch, Z.J. Xu, Y. Shao-Horn, Recommended practices and benchmark activity for hydrogen and oxygen electrocatalysis in water splitting and fuel cells, Adv. Mater., 31(31) (2019) 1806296. https://doi.org/10.1002/ADMA.201806296.
13. J.H. Montoya, L.C. Seitz, P. Chakthranont, A. Vojvodic, T.F. Jaramillo, J.K. Nørskov, Materials for solar fuels and chemicals, Nat. Mater., 16(1) (2017) 70-81. https://doi.org/10.1038/nmat4778.

14. J. Kibsgaard, I. Chorkendorff, Considerations for the scaling-up of water splitting catalysts, Nature Energy, 4(6) (2019) 430-433. https://doi.org/10.1038/s41560-019-0407-1.
15. M.F. Lagadec, A. Grimaud, Water electrolysers with closed and open electrochemical systems, Nat. Mater., 19(11) (2020) 1140-1150. https://doi.org/10.1038/s41563-020-0788-3.
16. M.D. Bhatt, J.Y. Lee, Advancement of platinum (Pt)-free (non-Pt precious metals) and/or metal-free (non-precious-metals) electrocatalysts in energy applications: A review and perspectives, Energy and Fuels, 34(6) (2020) 6634-6695. https://doi.org/10.1021/ACS.ENERGYFUELS.0C00953/ASSET/IMAGES/MEDIUM/EF0C00953_0026.GIF.
17. J. Yu, Y. Dai, Q. He, D. Zhao, Z. Shao, M. Ni, A mini-review of noble-metal-free electrocatalysts for overall water splitting in non-alkaline electrolytes, Materials Reports: Energy, 1(2) (2021) 100024. https://doi.org/10.1016/J.MATRE.2021.100024.
18. X. Xu, C. Su, Z. Shao, Fundamental understanding and application of $Ba_{0.5}Sr_{0.5}Co_{0.8}Fe_{0.2}O_{3-\delta}$perovskite in energy storage and conversion: past, present, and future, Energy and Fuels, 35(17) (2021) 13585-13609. https://doi.org/10.1021/ACS.ENERGYFUELS.1C02111/ASSET/IMAGES/LARGE/EF1C02111_0014.JPEG.
19. H. Sun, J. Dai, W. Zhou, Z. Shao, Emerging strategies for developing high-performance perovskite-based materials for electrochemical water splitting, Energy and Fuels, 34(9), (2020) 10547-10567. https://doi.org/10.1021/ACS.ENERGYFUELS.0C02313/ASSET/IMAGES/LARGE/EF0C02313_0010.JPEG.
20. B. Zhao, L. Zhang, D. Zhen, S. Yoo, Y. Ding, D. Chen, Y. Chen, Q. Zhang, B. Doyle, X. Xiong, M. Liu, A tailored double perovskite nanofiber catalyst enables ultrafast oxygen evolution, Nat. Commun., 8(1) (2017) 1-9. https://doi.org/10.1038/ncomms14586.
21. A. Li, S. Kong, C. Guo, H. Ooka, K. Adachi, D. Hashizume, Q. Jiang, H. Han, J. Xiao, R. Nakamura, Enhancing the stability of cobalt spinel oxide towards sustainable oxygen evolution in acid, Nature Catal., 5(2) (2022) 109-118. https://doi.org/10.1038/s41929-021-00732-9.
22. J. Zhang, J. Qian, J. Ran, P. Xi, L. Yang, D. Gao, Engineering lower coordination atoms onto NiO/Co_3O_4 heterointerfaces for boosting oxygen evolution reactions, ACS Catal., 10(21) (2020) 12376-12384. https://doi.org/10.1021/ACSCATAL.0C03756/ASSET/IMAGES/LARGE/CS0C03756_0006.JPEG.
23. D. Li, H. Liu, L. Feng, A review on advanced FeNi-based catalysts for water splitting reaction, Energy and Fuels, 34(11) (2020) 13491-13522. https://doi.org/10.1021/ACS.ENERGYFUELS.0C03084/ASSET/IMAGES/LARGE/EF0C03084_0015.JPEG.
24. J. Mohammed-Ibrahim, A review on NiFe-based electrocatalysts for efficient alkaline oxygen evolution reaction, J. Power Sources, 448 (2020) 227375. https://doi.org/10.1016/J.JPOWSOUR.2019.227375.
25. T.F. Jaramillo, K.P. Jørgensen, J. Bonde, J.H. Nielsen, S. Horch, I. Chorkendorff, Identification of active edge sites for electrochemical H_2 evolution from MoS_2 nanocatalysts, Science (1979) 317(5834) (2007) 100-102. https://doi.org/10.1126/SCIENCE.1141483/SUPPL_FILE/JARAMILLO_SOM.PDF.

26. B. Mohanty, A. Mitra, B. Jena, B.K. Jena, MoS_2 quantum dots as efficient electrocatalyst for hydrogen evolution reaction over a wide PH range, Energy and Fuels, 34(8) (2020) 10268-10275. https://doi.org/10.1021/ACS.ENERGYFUELS.0C01283/ASSET/IMAGES/LARGE/EF0C01283_0005.JPEG.
27. J. Zhang, Y. Chen, M. Liu, K. Du, Y. Zhou, Y. Li, Z. Wang, J. Zhang, 1T@2H-$MoSe_2$ nanosheets directly arrayed on Ti plate: An efficient electrocatalytic electrode for hydrogen evolution reaction, Nano Res., 11(9) (2018) 4587-4598. https://doi.org/10.1007/S12274-018-2040-X/METRICS.
28. P. Xiao, M.A. Sk, L. Thia, X. Ge, R.J. Lim, J.Y. Wang, K.H. Lim, X. Wang, Molybdenum phosphide as an efficient electrocatalyst for the hydrogen evolution reaction, Energy Environ. Sci., 7(8) (2014) 2624-2629. https://doi.org/10.1039/C4EE00957F.
29. W. Ahn, M.G. Park, D.U. Lee, M.H. Seo, G. Jiang, Z.P. Cano, F.M. Hassan, Z. Chen, Hollow multivoid nanocuboids derived from ternary Ni-Co-Fe Prussian Blue analog for dual-electrocatalysis of oxygen and hydrogen evolution reactions, Adv. Funct. Mater., 28(28) (2018) 1802129. https://doi.org/10.1002/ADFM.201802129.
30. K. Xu, P. Chen, X. Li, Y. Tong, H. Ding, X. Wu, W. Chu, Z. Peng, C. Wu, Y. Xie, Metallic nickel nitride nanosheets realizing enhanced electrochemical water oxidation, J. Am. Chem. Soc., 137(12) (2015) 4119-4125. https://doi.org/10.1021/JA5119495/SUPPL_FILE/JA5119495_SI_001.PDF.
31. B. Zhao, L. Zhang, D. Zhen, S. Yoo, Y. Ding, D. Chen, Y. Chen, Q. Zhang, B. Doyle, X. Xiong, M. Liu, A tailored double perovskite nanofiber catalyst enables ultrafast oxygen evolution, Nat. Commun., 8(1) (2017) 1-9. https://doi.org/10.1038/ncomms14586.
32. Q. Wang, Y. Lei, D. Wang, Y. Li, Defect engineering in earth-abundant electrocatalysts for CO_2 and N_2 reduction, Energy Environ. Sci., 12(6) (2019) 1730-1750. https://doi.org/10.1039/C8EE03781G.
33. J. Xiong, J. Di, J. Xia, W. Zhu, H. Li, J. Xiong, J. Di, J. Xia, W. Zhu, H. Li, Surface defect engineering in 2D nanomaterials for photocatalysis, Adv. Funct. Mater., 28(39) (2018) 1801983. https://doi.org/10.1002/ADFM.201801983.
34. Y. Liu, C. Xiao, Z. Li, Y. Xie, Vacancy engineering for tuning electron and phonon structures of two-dimensional materials, Adv. Energy Mater., 6(23) (2016) 1600436. https://doi.org/10.1002/AENM.201600436.
35. H. Arandiyan, S. Mofarah, C.C. Sorrell, E. Doustkhah, B. Sajjadi, D. Hao, Y. Wang, H. Sun, B.J. Ni, M. Rezaei, Z. Shao, T. Maschmeyer, Defect engineering of oxide perovskites for catalysis and energy storage: Synthesis of chemistry and materials science, Chem. Soc. Rev., 50(18) (2021) 10116-10211. https://doi.org/10.1039/D0CS00639D.
36. Y. Zheng, T.J. Slade, L. Hu, X.Y. Tan, Y. Luo, Z.Z. Luo, J. Xu, Q. Yan, M.G. Kanatzidis, Defect engineering in thermoelectric materials: What have we learned? Chem. Soc. Rev., 50(16) (2021) 9022-9054. https://doi.org/10.1039/D1CS00347J.
37. P. Babar, J. Mahmood, R.V. Maligal-Ganesh, S. Kim, Z. Xue, C.T. Yavuz, Electronic structure engineering for electrochemical water oxidation, J. Mater. Chem. A, 10 (2022), 20218-20241.
38. D. Nam, G. Lee, J. Kim, Effect of phosphorus vacancies on activity of Fe-doped nickel phosphide by $NaBH_4$ reduction for efficient oxygen evolution under alkaline conditions, J. Ind. Eng. Chem., 123 (2023) 201-208. https://doi.org/10.1016/J.JIEC.2023.03.035.

39. X. Li, Y. Liu, Q. Sun, Z. Huangfu, W.H. Huang, Z. Wang, C.C. Chueh, C.L. Chen, Z. Zhu, Effects of cationic and anionic defects on NiFe LDH in electrocatalytic oxygen evolution, ACS Sustain. Chem. Eng., 10(44) (2022) 14474-14485. https://doi.org/10.1021/ACSSUSCHEMENG.2C03887/ASSET/IMAGES/LARGE/SC2C03887_0007.JPEG.
40. H. Xue, A. Meng, T. Yang, Z. Li, C. Chen, Controllable oxygen vacancies and morphology engineering: Ultra-high HER/OER activity under base–acid conditions and outstanding antibacterial properties, J. Energy Chem., 71 (2022) 639-651. https://doi.org/10.1016/J.JECHEM.2022.04.052.
41. J. Liu, M. Zheng, M., J. Li, Y. Yuan, C. Li, S. Zhang, L. Yang, Z. Bai, J. Lu, Lithiation-induced defect engineering to promote oxygen evolution reaction, Adv. Funct. Mater., 33(2) (2023) 2209753. https://doi.org/10.1002/ADFM.202209753.
42. G. Yasin, S. Ibraheem, S. Ali, M. Arif, S. Ibrahim, R. Iqbal, A. Kumar, M. Tabish, M.A. Mushtaq, A. Saad, H. Xu, W. Zhao, Defects-engineered tailoring of tri-doped interlinked metal-free bifunctional catalyst with lower gibbs free energy of OER/HER intermediates for overall water splitting, Mater. Today Chem., 23 (2022) 100634. https://doi.org/10.1016/J.MTCHEM.2021.100634.
43. J. Han, X. Niu, J. Guan, Unveiling the role of defects in iron oxyhydroxide for oxygen evolution, J. Colloid Interface Sci., 635 (2023) 167-175. https://doi.org/10.1016/J.JCIS.2022.12.128.
44. K.P. Shejale, Y. Krishnan, R.K. Dharman, Y. Uk Jeong, S. Yeol Kim, Ultralow Fe instigated defect engineering of hierarchical N–porous carbon for highly efficient electrocatalysis, Mater. Des., 227 (2023) 111782. https://doi.org/10.1016/J.MATDES.2023.111782.
45. W. Deeloed, T. Priamushko, II. Jakub, C. Íž, S. Suramitr, F. Kleitz, Defect-engineered hydroxylated mesoporous spinel oxides as bifunctional electrocatalysts for oxygen reduction and evolution reactions, ACS Appl. Mater. Interfaces, 2330 (2022) 23307-23321. https://doi.org/10.1021/acsami.2c00254.
46. S. Guo, X. Wang, X. Zhou, H. Li, X. Ding, Tuning oxygen vacancies in Co_3O_4 nanorods through solvent reduction method for enhanced oxygen evolution activity, Energy and Fuels, 37(7) (2023) 5421-5428. https://doi.org/10.1021/ACS.ENERGYFUELS.3C00034/ASSET/IMAGES/LARGE/EF3C00034_0010.JPEG.
47. S.F. Li, J. Zheng, L. Hu, Y. Ma, D. Yan, Facile surface defect engineering on perovskite oxides for enhanced OER performance, Dalton Transactions, 52(13) (2023) 4207-4213. https://doi.org/10.1039/D2DT04019K.
48. L. Bo, W. Shi, F. Nian, Y. Hu, L. Pu, P. Li, Z. Zhang, J. Tong, Interface engineering of Co_3S_4@Co_3O_4/N, S-doped carbon Core@shell nanostructures serve as an excellent bifunctional ORR/OER electrocatalyst for rechargeable Zn-air battery, Sep Purif Technol., 307 (2023) 122536. https://doi.org/10.1016/J.SEPPUR.2022.122536.
49. F. Xu, X. Liu, L. Zhang, M. Guo, M. Li, X. Ding, L. Zhang, Revealing and optimizing the dialectical relationship between NOR and OER: Cation vacancy engineering enables RuO_2 with unanticipated high electrochemical nitrogen oxidation performance, Adv. Energy Mater., 13(22) (2023) 2300615. https://doi.org/10.1002/AENM.202300615.
50. C. Wang, D. Liu, K. Zhang, H. Xu, R. Yu, X. Wang, Y. Du, Defect and interface engineering of three-dimensional open nanonetcage electrocatalysts for advanced electrocatalytic oxygen evolution reaction, ACS Appl. Mater. Interfaces, 14 (2022)

38669-38676. https://doi.org/10.1021/ACSAMI.2C07792/ASSET/IMAGES/LARGE/AM2C07792_0006.JPEG.
51. P. Adiga, W. Nunn, C. Wong, A.K. Manjeshwar, S. Nair, B. Jalan, K.A. Stoerzinger, Breaking OER and CER scaling relations via strain and its relaxation in RuO_2 (101), Mater. Today Energy, 28 (2022) 101087. https://doi.org/10.1016/J.MTENER.2022.101087.
52. J. Song, S. Zhao, D. Liu, Y. Xiong, F. Hu, L. Li, L. Li, H. Pan, S. Peng, Plasma-induced defect engineering of porous metal–organic framework nanosheet arrays for efficient water splitting, Chem. Commun., 58(69) (2022) 9662-9665. https://doi.org/10.1039/D2CC03633A.
53. H. Yan, Z. Jiang, B. Deng, Y. Wang, Z.J. Jiang, Ultrathin carbon coating and defect engineering promote RuO_2 as an efficient catalyst for acidic oxygen evolution reaction with super-high durability, Adv. Energy Mater., 13(23) (2023) 2300152. https://doi.org/10.1002/AENM.202300152.
54. H. Xu, C. Wang, G. He, H. Chen, Engineering defective trimetallic metal–organic framework nanosheets for advanced water oxidation electrocatalysis, Dalton Transactions, 52(24) (2023) 8466-8472. https://doi.org/10.1039/D3DT01228J.
55. H.A. Alburaih, S. Manzoor, M. Abdullah, M.N. Ashiq, S. Aman, S.V. Trukhanov, T.I. Zubar, Z. Sun, T.A. Taha, A.V. Trukhanov, Electro-oxidation reconstitution of aluminium copper MOF-derived metal oxyhydroxides for a robust OER process, RSC Adv., 13 (2023) 8736-8742.
56. R. Li, H. Qi, F. Pan, W. Xie, L. Zhou, H. Liu, J. Hu, Defect engineering in bimetallic NiFe-BTC for boosting electrocatalytic oxygen evolution reaction through coordinated ionic liquids, ChemElectroChem, 10(13) (2023) e202300103. https://doi.org/10.1002/CELC.202300103.
57. L. Xia, L. Bo, W. Shi, Y. Zhang, Y. Shen, X. Ji, X. Guan, Y. Wang, J. Tong, Defect and interface engineering of templated synthesis of hollow porous $Co_3O_4/CoMoO_4$ with highly enhanced electrocatalytic activity for oxygen evolution reaction, J. Chem. Eng. 452 (2023) 139250. https://doi.org/10.1016/J.CEJ.2022.139250.
58. Y. Hao, S.F. Hung, W.J. Zeng, Y. Wang, C. Zhang, C.H. Kuo, L. Wang, S. Zhao, Y. Zhang, H.Y. Chen, S. Peng, Switching the oxygen evolution mechanism on atomically dispersed Ru for enhanced acidic reaction kinetics, J. Am. Chem. Soc., 145(43) (2023) 23659-23669. https://doi.org/10.1021/JACS.3C07777/ASSET/IMAGES/LARGE/JA3C07777_0006.JPEG.
59. Z. Li, C. Zhang, Y. Yang, S. Pi, Y. Yu, C. Wan, B. Zhou, W. Chao, L. Lu, Molten-salt-induced phosphorus vacancy defect engineering of heterostructured cobalt phosphides for efficient overall water splitting, Inorg. Chem. Front., 10(1) (2022) 325-334. https://doi.org/10.1039/D2QI01902G.
60. Y. Hu, J. Zhou, L. Li, Z. Hu, T. Yuan, C. Jing, R. Liu, S. Xi, H. Jiang, J.Q. Wang, L. Zhang, Dynamic structural transformation induced by defects in nano-rod FeOOH during electrochemical water splitting, J. Mater. Chem. A. Mater., 10(2) (2022) 602-610. https://doi.org/10.1039/D1TA08938B.
61. X. Luo, L. Zhang, M. Guo, Z. Liu, D. Wu, D. Zhen, Y. Liu, Engineering the structural defects of spinel oxide nanoneedles by doping of v for a highly efficient oxygen evolution reaction, ACS Appl. Mater. Interfaces, 14 (2022) 50055-50067. https://doi.org/10.1021/ACSAMI.2C15524/ASSET/IMAGES/LARGE/AM2C 15524_0009.JPEG.

62. Y. Zhao, Q. Wen, D. Huang, C. Jiao, Y. Liu, J. Fang, M. Sun, L. Yu, Y. Zhao, Operando reconstruction toward dual-cation-defects co-containing NiFe oxyhydroxide for ultralow energy consumption industrial water splitting electrolyzer, Adv. Energy Mater., 13(10) (2023) 2203595. https://doi.org/10.1002/AENM.202203595.
63. A. Mahmood, G. Lu, X. Wang, Y. Wang, X. Xie, J. Sun, Investigating the stability and role of defects in vertically aligned WS_2/MoS_2 heterojunctions on OER activity using first principles study, J. Power Sources, 551 (2022) 232208. https://doi.org/10.1016/J.JPOWSOUR.2022.232208.
64. J. Yao, Y. Ji, L. Pei, S. Tan, F. Ren, Amorphous FeOOH nanoparticles decorated on defect-rich porous Ni MOF nanosheet based hierarchical architectures toward superior OER performance, New J. Chem., 46(20) (2022) 9650-9657. https://doi.org/10.1039/D2NJ01385A.
65. X. Li, K. Zheng, J. Zhang, G. Li, C. Xu, Engineering sulfur vacancies in spinel-phase Co_3S_4 for effective electrocatalysis of the oxygen evolution reaction, ACS Omega, 7(14) (2022) 12430-12441. https://doi.org/10.1021/ACSOMEGA.2C01423/ASSET/IMAGES/LARGE/AO2C01423_0009.JPEG.
66. Q. Wang, H. Ma, X. Ren, X. Sun, X. Liu, D. Wu, Q. Wei, Defect engineering and atomic doping of porous Co-Ni 2P nanosheet arrays for boosting electrocatalytic oxygen evolution, Nanoscale Adv., 5(14) (2023) 3691-3696. https://doi.org/10.1039/D3NA00217A.
67. Q. Ouyang, S. Cheng, C. Yang, Z. Lei, Ni, Co, and Yb cation co-doping and defect engineering of FeOOH nanorods as an electrocatalyst for the oxygen evolution reaction, Inorg. Chem., 62(4) (2023) 1719-1727. https://doi.org/10.1021/ACS.INORGCHEM.2C04174/ASSET/IMAGES/LARGE/IC2C04174_0006.JPEG.
68. X. Zhang, Y. Yang, Y. Liu, Z. Jia, Q. Wang, L. Sun, L.C. Zhang, J.J. Kruzic, J. Lu, B. Shen, Defect engineering of a high-entropy metallic glass surface for high-performance overall water splitting at ampere-level current densities, Adv. Mater., 35(38) (2023) 2303439. https://doi.org/10.1002/ADMA.202303439.
69. L. Wu, M. Ning, X. Xing, Y. Wang, F. Zhang, G. Gao, S. Song, D. Wang, C. Yuan, L. Yu, J. Bao, S. Chen, Z. Ren, Boosting oxygen evolution reaction of (Fe, Ni) OOH via defect engineering for anion exchange membrane water electrolysis under industrial conditions, Adv. Mater., 35(44) (2023) 2306097. https://doi.org/10.1002/ADMA.202306097.
70. Y. Cao, L. Li, X., Yu, M. Tahir, Z. Xiang, W. Kong, Z. Lu, X. Xing, Y. Song, Engineering vacancies at the 2D nanocrystals for robust bifunctional electrocatalysts, ACS Appl. Mater. Interfaces, 14 (2022) 56734. https://doi.org/10.1021/ACSAMI.2C15955/ASSET/IMAGES/LARGE/AM2C15955_0006.JPEG.
71. J. Li, F. Yang, Y. Du, M. Jiang, X. Cai, Q. Hu, J. Zhang, The critical role of A, B-site cations and oxygen vacancies on the OER electrocatalytic performances of $Bi_{0.15}Sr_{0.85}Co_{1-x}Fe_xO_{3-\delta}$ ($0.2 \leq x \leq 1$) perovskites in alkaline media, J. Chem. Eng., 451 (2023) 138646. https://doi.org/10.1016/J.CEJ.2022.138646.
72. L. Wang, J. Huang, Z. Huang, H. Li, T. Taylor Isimjan, X. Yang, Revealing dynamic structural evolution of V and P co-doping-induced Co defects as large-current water oxidation catalyst, J. Chem. Eng., 472 (2023) 144924. https://doi.org/10.1016/J.CEJ.2023.144924.
73. L. Tian, Z. Chen, T. Wang, M. Cao, X. Lu, W. Cheng, C. He, J. Wang, Z. Li, Mo doping and Se vacancy engineering for boosting electrocatalytic water oxidation by regulating the electronic structure of self-supported Co_9 Se_8 @NiSe, Nanoscale, 15(1) (2022) 259-265. https://doi.org/10.1039/D2NR05410H.

74. D. Nam, E. Jang, J. Kim, Phosphorus doping and sulfur vacancies defect engineering for efficient electrocatalytic water-splitting by modulating the electronic structure of the cobalt sulfide with synergy effect, J. Alloys Compd., 947 (2023) 169625. https://doi.org/10.1016/J.JALLCOM.2023.169625.
75. B. Zhang, J. Shan, X. Wang, Y. Hu, Y. Li, B. Zhang, J. Shan, X. Wang, Y. Hu, Y. Li, Ru/Rh cation doping and oxygen-vacancy engineering of FeOOH nanoarrays@$Ti_3C_2T_x$ MXene heterojunction for highly efficient and stable electrocatalytic oxygen evolution, Small, 18(25) (2022) 2200173. https://doi.org/10.1002/SMLL.202200173.
76. K. Chen, X. Wang, C. Zhang, R. Xu, H. Wang, L. Chu, M. Huang, Three-phases $Co/Co_9S_8/MnS$ heterostructures engineering for boosted ORR/OER activities in Zn–air batteries, Mater. Today Energy, 30 (2022) 101150. https://doi.org/10.1016/J.MTENER.2022.101150.
77. S. Liu, H. Zhang, E. Hu, T. Zhu, C. Zhou, Y. Huang, M. Ling, X. Gao, Z. Lin, Boosting oxygen evolution activity of NiFe-LDH using oxygen vacancies and morphological engineering, J. Mater. Chem. A Mater., 9(41) (2021) 23697-23702. https://doi.org/10.1039/D1TA06263H.
78. Z. Luo, R. Miao, T. Doan Huan, I.M. Mosa, A.S. Poyraz, W. Zhong, J.E. Cloud, D.A. Kriz, S. Thanneeru, J. He, Y. Zhang, R. Ramprasad, S.L. Suib, Z. Luo, T.D. Huan, W. Zhong, J.E. Cloud, J. He, R. Miao, Mesoporous MoO_{3-x} material as an efficient electrocatalyst for hydrogen evolution reactions, Adv. Energy Mater., 6(16) (2016) 1600528. https://doi.org/10.1002/AENM.201600528.
79. R. Wu, J. Zhang, Y. Shi, D. Liu, B. Zhang, Metallic WO_2-carbon mesoporous nanowires as highly efficient electrocatalysts for hydrogen evolution reaction, J. Am. Chem. Soc., 137(22) (2015) 6983-6986. https://doi.org/10.1021/JACS.5B01330/SUPPL_FILE/JA5B01330_SI_001.PDF.
80. Z. Xiao, C. Xie, Y. Wang, R. Chen, S. Wang, Recent advances in defect electrocatalysts: Preparation and characterization, J. Energy Chem., 53 (2021) 208-225. https://doi.org/10.1016/J.JECHEM.2020.04.063.
81. C. Tsai, H. Li, S. Park, J. Park, H.S. Han, J.K. Nørskov, X. Zheng, F. Abild-Pedersen, Electrochemical generation of sulfur vacancies in the basal plane of MoS_2 for hydrogen evolution, Nat. Commun., 8(1) (2017) 1-8. https://doi.org/10.1038/ncomms15113.
82. H. Li, C. Tsai, A.L. Koh, L. Cai, A.W. Contryman, A.H. Fragapane, J. Zhao, H.S. Han, H.C. Manoharan, F. Abild-Pedersen, J.K. Nørskov, X. Zheng, Activating and optimizing MoS_2 basal planes for hydrogen evolution through the formation of strained sulphur vacancies, Nat. Mater., 15(1) (2015) 48-53. https://doi.org/10.1038/nmat4465.
83. Q. Xie, Z. Cai, P. Li, D. Zhou, Y. Bi, X. Xiong, E. Hu, Y. Li, Y. Kuang, X. Sun, Layered double hydroxides with atomic-scale defects for superior electrocatalysis, Nano Res., 11(9) (2018) 4524-4534. https://doi.org/10.1007/S12274-018-2033-9/METRICS.
84. C. Su, W. Wang, Z. Shao, Cation-deficient perovskites for clean energy conversion, Acc Mater. Res., 2(7) (2021) 477-488. https://doi.org/10.1021/ACCOUNTSMR.1C00036/ASSET/IMAGES/LARGE/MR1C00036_0003.JPEG.
85. Y.S. Xie, Z. Wang, M. Ju, X. Long, S. Yang, Dispersing transition metal vacancies in layered double hydroxides by ionic reductive complexation extraction for efficient water oxidation, Chem. Sci., 10(36) (2019) 8354-8359. https://doi.org/10.1039/C9SC02723H.

86. T. Chen, S. Qin, M. Qian, H. Dai, Y. Fu, Y. Zhang, B. Ye, Q. Lin, Q. Yang, Defect-rich Fe-doped CoP nanosheets as efficient oxygen evolution electrocatalysts, Energy and Fuels, 35(13) (2021) 10890-10897. https://doi.org/10.1021/ACS.ENERGYFUELS.1C01301/ASSET/IMAGES/LARGE/EF1C01301_0006.JPEG.
87. J. Zhang, Y. Wang, J. Cui, J. Wu, Y. Li, T. Zhu, H. Kang, J. Yang, J. Sun, Y. Qin, Y. Zhang, P.M. Ajayan, Y. Wu, Water-soluble defect-rich MoS_2 ultrathin nanosheets for enhanced hydrogen evolution, J. Phys. Chem. Lett., 10(12) (2019) 3282-3289. https://doi.org/10.1021/ACS.JPCLETT.9B01121/ASSET/IMAGES/LARGE/JZ-2019-01121F_0005.JPEG.
88. X. Tong, Y. Zhao, Z. Zhuo, Z. Yang, S. Wang, Y. Liu, N. Lu, H. Li, T. Zhai, Dual-regulation of defect sites and vertical conduction by spiral domain for electrocatalytic hydrogen evolution, Angewandte Chemie, 134(7) (2022) e202112953. https://doi.org/10.1002/ANGE.202112953.
89. X. Li, Q. Hu, H. Wang, M. Chen, X. Hao, Y. Ma, J. Liu, K. Tang, A. Abudula, G. Guan, Charge induced crystal distortion and morphology remodeling: Formation of Mn-CoP nanowire@ Mn-CoOOH nanosheet electrocatalyst with rich edge dislocation defects, Appl. Catal. B, 292 (2021) 120172. https://doi.org/10.1016/J.APCATB.2021.120172.
90. X. Zhang, Y. Liu, J. Gao, G. Han, M. Hu, X. Wu, H. Cao, X. Wang, B. Li, Defect-rich $(Co–CoS_2)_x$ @Co_9S_8 nanosheets derived from monomolecular precursor pyrolysis with excellent catalytic activity for hydrogen evolution reaction, J. Mater. Chem. A Mater., 6(17) (2018) 7977-7987. https://doi.org/10.1039/C8TA01163J.
91. W. Dai, Y. Zhu, Y. Ye, Y. Pan, T. Lu, S. Huang, Electrochemical incorporation of heteroatom into surface reconstruction induced Ni vacancy of NixO nanosheet for enhanced water oxidation, J. Colloid Interface Sci., 608 (2022) 3030-3039. https://doi.org/10.1016/J.JCIS.2021.11.026.
92. Z. Xiao, W. Zhou, N. Zhang, C. Liao, S. Huang, G. Chen, G. Chen, M. Liu, X. Liu, R. Ma, Lithium doped nickel oxide nanocrystals with a tuned electronic structure for oxygen evolution reaction, Chem. Commun., 57(49) (2021) 6070-6073. https://doi.org/10.1039/D1CC01655E.
93. J. Ran, L. Wang, M. Si, X. Liang, D. Gao, J. Ran, L. Wang, D. Gao, M. Si, X. Liang, Tailoring spin state of perovskite oxides by fluorine atom doping for efficient oxygen electrocatalysis, Small, 19(10) (2023) 2206367. https://doi.org/10.1002/SMLL.202206367.
94. Y. Huang, M. Li, F. Pan, Z. Zhu, H. Sun, Y. Tang, G. Fu, Plasma-induced Mo-doped Co_3O_4 with enriched oxygen vacancies for electrocatalytic oxygen evolution in water splitting, Carbon Energy, 5(3) (2023) e279. https://doi.org/10.1002/CEY2.279.
95. D. Yan, Y. Li, J. Huo, R. Chen, L. Dai, S. Wang, D. Yan, Y. Li, J. Huo, R. Chen, S. Wang, L. Dai, Defect chemistry of nonprecious-metal electrocatalysts for oxygen reactions, Adv. Mater., 29(48) (2017) 1606459. https://doi.org/10.1002/ADMA.201606459.
96. D. Wang, Q. Li, C. Han, Q. Lu, Z. Xing, X. Yang, Atomic and electronic modulation of self-supported nickel-vanadium layered double hydroxide to accelerate water splitting kinetics, Nat. Commun., 10(1) (2019) 1-12. https://doi.org/10.1038/s41467-019-11765-x.
97. M.W. Louie, A.T. Bell, An investigation of thin-film Ni-Fe oxide catalysts for the electrochemical evolution of oxygen, J. Am. Chem. Soc., 135(33) (2013) 12329-12337. https://doi.org/10.1021/JA405351S/SUPPL_FILE/JA405351S_SI_001.PDF.

98. M. Asnavandi, Y. Yin, Y. Li, C. Sun, C. Zhao, Promoting oxygen evolution reactions through introduction of oxygen vacancies to benchmark NiFe-OOH catalysts, ACS Energy Lett., 3(7) (2018) 1515-1520. https://doi.org/10.1021/ACSENERGYLETT.8B00696/ASSET/IMAGES/LARGE/NZ-2018-00696N_0003.JPEG.
99. L. Ma, S. Chen, Z. Pei, H. Li, Z. Wang, Z. Liu, Z. Tang, J.A. Zapien, C. Zhi, Flexible waterproof rechargeable hybrid zinc batteries initiated by multifunctional oxygen vacancies-rich cobalt oxide, ACS Nano, 12(8) (2018) 8597-8605. https://doi.org/10.1021/ACSNANO.8B04317/ASSET/IMAGES/MEDIUM/NN-2018-043174_M002.GIF.
100. J. Bao, X. Zhang, B. Fan, J. Zhang, M. Zhou, W. Yang, X. Hu, H. Wang, B. Pan, Y. Xie, Ultrathin spinel-structured nanosheets rich in oxygen deficiencies for enhanced electrocatalytic water oxidation, Angew. Chem. Int. Ed. 54(25) (2015) 7399-7404. https://doi.org/10.1002/ANIE.201502226.
101. Z. Cai, Y. Bi, E. Hu, W. Liu, N. Dwarica, Y. Tian, X. Li, Y. Kuang, Y. Li, X.Q. Yang, H. Wang, X. Sun, Single-crystalline ultrathin Co_3O_4 nanosheets with massive vacancy defects for enhanced electrocatalysis, Adv. Energy Mater., 8(3) (2018) 1701694. https://doi.org/10.1002/AENM.201701694.
102. Z. Xiao, Y. Wang, Y.C. Huang, Z. Wei, C.L. Dong, J. Ma, S. Shen, Y. Li, S. Wang, Filling the oxygen vacancies in Co_3O_4 with phosphorus: An ultra-efficient electrocatalyst for overall water splitting, Energy Environ. Sci., 10(12) (2017) 2563-2569. https://doi.org/10.1039/C7EE01917C.
103. R. Liu, Y. Wang, D. Liu, Y. Zou, S. Wang, R. Liu, Y. Wang, D. Liu, S. Wang, Y. Zou, Water-plasma-enabled exfoliation of ultrathin layered double hydroxide nanosheets with multivacancies for water oxidation, Adv. Mater., 29(30) (2017) 1701546. https://doi.org/10.1002/ADMA.201701546.
104. H.J. Lee, S. Back, J.H. Lee, S.H. Choi, Y. Jung, J.W. Choi, Mixed transition metal oxide with vacancy-induced lattice distortion for enhanced catalytic activity of oxygen evolution reaction, ACS Catal., 9(8) (2019) 7099-7108. https://doi.org/10.1021/ACSCATAL.9B01298/ASSET/IMAGES/LARGE/CS-2019-012983_0007.JPEG.
105. Z.L. Zhao, Q. Wang, X. Huang, Q. Feng, S. Gu, Z. Zhang, H. Xu, L. Zeng, M. Gu, H. Li, Boosting the oxygen evolution reaction using defect-rich ultra-thin ruthenium oxide nanosheets in acidic media, Energy Environ. Sci., 13(12) (2020) 5143-5151. https://doi.org/10.1039/D0EE01960G.
106. X. Li, X. Su, Y. Pei, J. Liu, X. Zheng, K. Tang, G. Guan, X. Hao, Generation of edge dislocation defects in Co_3O_4 catalysts: An efficient tactic to improve catalytic activity for oxygen evolution, J. Mater. Chem. A Mater., 7(17) (2019) 10745-10750. https://doi.org/10.1039/C9TA01697J.
107. Y.S. Xie, Z. Wang, M. Ju, X. Long, S. Yang, Dispersing transition metal vacancies in layered double hydroxides by ionic reductive complexation extraction for efficient water oxidation, Chem. Sci., 10(36) (2019) 8354-8359. https://doi.org/10.1039/C9SC02723H.

CHAPTER
4

Beyond Boundaries: Unleashing Applications of Hybrid Nanomaterials for Photodetectors

Mukesh Kumar Thakur*, Aswin Asaithambi, Luca Rebecchi, Sidharth Kuriyil and Ilka Kriegel*

Functional Nanosystems, Istituto Italiano di Tecnologia, via Morego 30, 16163 Genova, Italy

INTRODUCTION

1. What are the Nanomaterials?

Nanomaterials originate from the prefix "nano-," signifying the nanoscale and the term "materials." These substances undergo engineering and manipulation at the nanometer level, commonly falling within the size range of 1 to 100 nanometers. According to the International Organization for Standardization (ISO), the term "nano" refers to a size ranging from 1 to 100 nm, denoted as nano:1. In 1959, Richard Feynman delivered a talk at the annual meeting of the American Physical Society titled "There's Plenty of Room at the Bottom: An Invitation to Enter a New Field of Physics." Following his talk, the era of "Nanotechnology" began, and researchers began developing nanomaterials [1–3].

1.1 Nanomaterials

Nanomaterials are characterized by having at least one external dimension measuring 100 nanometers (nm) or less. This specific size range distinguishes nanomaterials, emphasizing their nanoscale dimensions that contribute to unique physical and chemical properties [3]. Despite sharing composition with their

*Corresponding author: mukesh.thakur@iit.it, and ilka.kriegel@iit.it

bulk counterparts, nanomaterials exhibit distinctive physicochemical properties. At the nanoscale, these materials bear unexpected characteristics that are absent at larger scales. This encompasses phenomena like the transition of opaque substances to transparency (e.g., copper), the catalytic activation of inert materials (e.g., platinum), the transformation of stable materials into combustible forms (e.g., aluminum), the conversion of solids to liquids at room temperature (e.g., gold), and the evolution from insulators to conductors (e.g., silicon) [4,5].

1.2 Classification of Nanomaterials

Nanomaterials are generally classified into four groups based on their dimensions and size: (a) Bulk (3D), (b) 2D, (c) 1D, and (d) 0D [4,6]. These nanomaterials exhibit specific characteristics. Notably, crucial properties like the density of states undergo significant changes, resulting in unprecedented properties in nanomaterials, including alterations in energy band diagrams and carrier confinement [7]. Figure 1 shows the schematic description of how nanotechnology opens new possibilities with various nanomaterials, and its potential hybrid materials open new avenues for various applications [2,3].

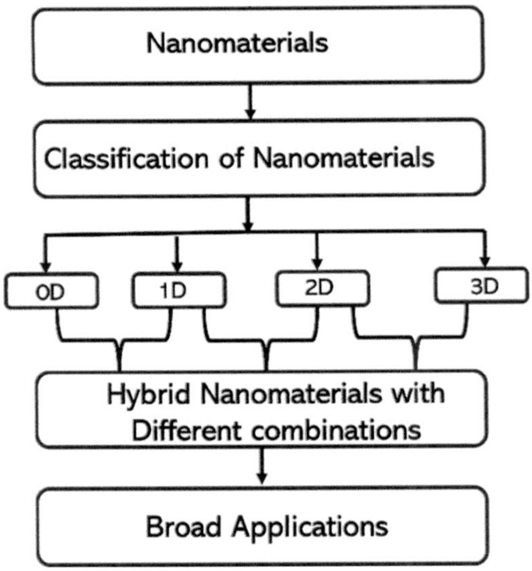

Figure 1. Schematic description of how nanotechnology opens new possibilities with various nanomaterials and its potential hybrid materials open new avenues for various applications.

1.3 0D Nanomaterials

Nanomaterials with all external dimensions at the nanoscale, between 1 and 100 nm, are known as zero-dimensional (0D) nanomaterials. Typically, 0D nanomaterials manifest as nanoparticles. The known 0D nanomaterials are quantum dots, fullerene, metallic nanoparticles, etc. Figure 2 shows a fullerene nanostructure as an example of 0D nanomaterials.

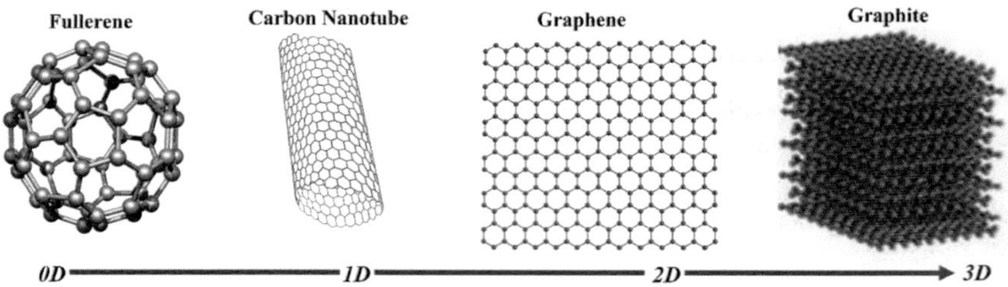

Figure 2. Examples of OD, 1D, 2D, and 3D materials.

1.4 One-Dimensional (1D) Nanomaterials

The second class of nanoscale materials exhibits confinement along two dimensions, hence being termed one-dimensional (1D) nanomaterials. Their typical size ranges between 50-300 nm in diameter or equivalent cross-sections [32]. 1D nanomaterials have garnered considerable attention from the scientific community due to their distinctive optical, physical, and chemical properties, offering a wide range of potential applications [32, 33]. Common examples of 1D nanomaterials include nanotubes (refer to Figure 2), nanofibers, nanowires, and nanorods.

1.5 Two-dimensional (2D) Nanomaterials

Two-dimensional (2D) materials have garnered significant attention within the scientific community in recent years due to their unique properties at the nanometer scale. Also known as layered materials, they encompass various forms, such as nanoribbons, nanoplates, nanowalls, and nanoflakes [83]. The quantum confinement effects in these materials result in distinctive physical, optical, and chemical properties, setting them apart from their bulk counterparts. Well-known examples of 2D materials include Graphene (Figure 2), transition metal dichalcogenides (TMDCs), black phosphorus, metal oxides, and many others.

1.6 Three-dimensional (3D) Materials

Three-dimensional (3D) materials refer to materials with significant dimensions in all three spatial directions: length, width, and height. In contrast to one-dimensional (1D) or two-dimensional (2D) materials, 3D materials exhibit significant size and volume across all dimensions, distinguishing them from their lower-dimensional counterparts. A typical example is shown in Figure 2. For example, most of the bulk materials (like graphite shown in Figure 2) are 3D Dimensional.

We have explained nanomaterials and types of nanomaterials above with a few examples, as shown in Figure 2. Each individual material always has very specific properties depending on its material but in particular also on its dimension.

1.7 Hybrid Nanomaterials

Hybrid nanocomposite materials are combinations of nanomaterials with different dimensional materials; for example, 0D with 1D, 2D, and 3D nanomaterials. These innovative nanocomposites typically exhibit superior physical and chemical properties in comparison to individual materials [4,6]. This characteristic has led to their diverse applications, including but not limited to energy storage devices, sensors, catalysis, flexible electronics, plasmonics, photodetectors, and numerous others [8]. In this chapter, our primary focus is on light detection devices known as photodetectors [9].

1.8 Photodetectors (PD)

A photodetector is a light sensor that converts an optical signal to an electrical signal. PDs play a crucial role in numerous devices and applications, including image sensors, electronics, spectrometry, biomedical imaging, etc. [10]. While PDs utilizing conventional semiconductor materials offer numerous advantages, they come with certain drawbacks related to manufacturing costs and integration processes, hindering their widespread application in consumer goods and defense. The basic steps involved in photodetection are as follows.

(a) **Generation of charge carriers in active materials:** When a photodetector is exposed to light (depending upon the absorption or band gap of materials), photons from the light source interact with the material (mostly semiconductors) in the detector. This interaction generates electron-hole pairs in the material. The energy from the absorbed photons is transferred to electrons, allowing them to move freely within the material. Modern photodetectors operate based on the internal photo effect (photoconductivity), where the photo-excited carriers remain within the sample [10].

(b) **Separation and collection of generated charge carriers:** The application of an electric field (voltage) initiates the movement of these carriers, resulting in the generation of a current. This current is then collected at the end of the metal electrode.

The performance of photodetectors can be assessed using the following parameters.

Photoresponsivity (R): It is defined as the ratio of the photocurrent generated by light (photocurrent, denoted as ΔI) to the incident power (denoted as P_{in}) of light on the active area of the photodetector device [11]

$$R = \frac{\Delta I}{P} \tag{1}$$

Photocurrent gain (η):

$$\eta = R\frac{hc}{e\lambda} \tag{2}$$

where h is Planck's constant, c is the velocity of light, e electronic charge and λ is the wavelength of light [11].

External quantum efficiency (EQE): The EQE is defined as the ratio of the number of charge carriers collected by the device to the number of incident photons with a specific energy [12].

Internal external quantum efficiency (IQE): The IQE is characterized by the ratio of the number of electron-hole pairs or charge carriers generated to the number of photons absorbed within the device's active layer [12].

Detectivity (D^*): The sensitivity of photodetectors is determined by an important parameter known as Detectivity [11]

$$D^* = \frac{R\sqrt{A}}{\sqrt{2eI_d}} \quad \text{or} \quad \frac{\sqrt{AB}}{NEP} \tag{3}$$

where R is the responsivity, 'A' is the active area of the device's, e is the electronic charge, and I_d is the dark current. D^* is strongly dependent on dark current. B is the bandwidth. which is inversely proportional to the response (τ) of the photodetector ($B = 1/\tau$), and NEP is the noise equivalent power.

NEP is defined as mathematically:

$$\text{NEP} = \frac{I_d}{R} \text{also } I_d^2 = 2eI_dB \tag{4}$$

The response time of the photodetector: The response time, indicated by rise (τ_{rise}) and fall (τ_{fall}) times, is a crucial parameter for a photodetector. The rise time (τ_{rise}) and fall time (τ_{fall}) represent the duration for ΔI to increase from 10% to 90% or decrease from 90%% to 10%% of its saturated value, respectively. The response time of a photodetector can vary from seconds to nanoseconds, depending on the materials used and the device structures employed. The detailed calculations can be found in the supplementary materials. This chapter explores the photodetectors based on 0D, 1D, 2D, and 3D structures and their hybrid counterparts, including the synthesis of hybrid materials, photodetection mechanism, and performance parameters.

1.9 Photodetectors Based on 0D Nanomaterials

The colloidal quantum dots (CQDs) are a well-known example of 0D materials. CQDs are crucial in nanoscale phenomena due to the quantum confinement effect [13]. These nanostructured materials, derived from a solution phase, exploit the quantum size effect, resulting in a bandgap strongly dependent on nanocrystal size [13]. The advancements in colloidal quantum dot synthesis and photophysics [13] allow precise control over size, shape, and composition [14]. and led to the Nobel Prize in Chemistry in 2024 [15]. The confinement of charge carriers within these quantum dots, akin to electrons in an atom, influences various charge transport mechanisms in quantum dot solids [16]. These mechanisms encompass delocalization [16], variable-range hopping [17], slow carrier relaxation [18] and interactions between electrons and holes [19–24]. Nanoparticles leverage multicarrier effects for enhanced performance in photodetection through carrier multiplication [15]. Moreover, these effects hold broader applications, providing opportunities for optical gain in lasing, facilitating strong nonlinear interactions in optical signal processing, and offering the potential for improved light harvesting in photovoltaics and photodetectors [24].

In Figure 3a, various sizes of quantum dots (QDs) are depicted, showcasing their optical and structural characteristics. The impact of quantum confinement is evident, as emphasized in the blue-shift observed in the absorption edge and emission peak of the QDs with decreasing size. This shift is visually represented

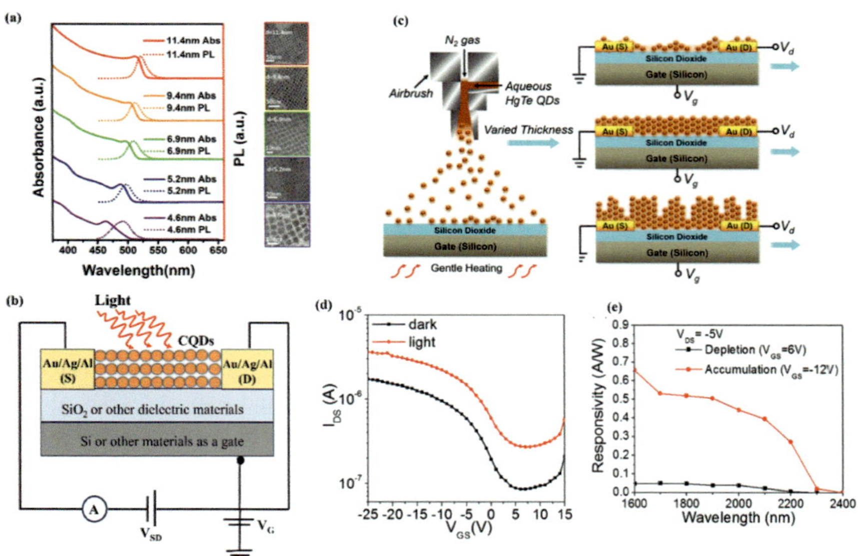

Figure 3. (a) The absorption and photoluminescence spectra (λ_{ex} = 430 nm) of $CsPbBr_3$ quantum dots (QDs) with varying size (left), and high-resolution transmission electron microscopy (TEM) images (right). (b) Schematic representation of a photodetector device based on CQDs. (c) Schematics of the spray-coating setup and structures of aqueous HgTe QDs-based phototransistor devices with varying QD film thicknesses deposited through a series of spray passes. (d) Typical transfer characteristics of the aqueous HgTe QD-based phototransistor (30 spray passes) under dark and illuminated conditions. The curves were measured in a forward scan with V_{GS} ranging from +15 V to −25 V. (e) Photocurrent dependence on light intensity and corresponding log-log fit for the aqueous HgTe QD-based phototransistor (30 spray passes). (Figure a reprinted with permission Chen et al. [25] Copyright © 2017, American Chemical Society, Figures c-e reprinted with permission Chen et al. [26] Copyright © 2017, American Chemical Society.)

in the absorption and photoluminescence spectra presented in Figure 3a (left) and 1a (right), where the diameters of the QDs range from 11.4 to 4.6 nm [25,27,28]. This demonstrates the ease of tuning optical absorption in CQDs [24]. Figure 3b illustrates the simple architecture of a CQD based phototransistor. The ease of spin-coating or drop-coating CQDs onto a chosen substrate allows for the creation of diverse layers of CQD structures, opening numerous applications such as solar cells and random lasers, photodetectors, and light emitting diode (LED), and many other fields [29,30]. For a simple example, Chen et al. [26] demonstrated a HgTe QDs-based photodetector, as shown in Figure 3c. The fully automated, aprotic solvent gas-injection synthesis facilitates the precise and controlled production of high-quality HgTe quantum dots (QDs). The layer-by-layer spray-coating process permits the optimization of the QD layer's thickness, thereby achieving a gating effect in the transistor structure at room temperature, as shown in Figure 3c. Subsequently, a bottom-contact/bottom-gate transistor (depicted in Figure 3c) is prepared, utilizing a heavily doped silicon substrate as the gate and its oxide surface layer (approximately 300 nm thick SiO_2) as the gate dielectric. Photolithographically patterned gold electrodes serve as the source and drain contacts. For the creation of the photoactive layer, the aqueous HgTe CQD solution is spray-deposited onto the substrate through multiple passes in an ambient environment. The I-V characteristics of the HgTe CQDs device are illustrated in Figure 3d, revealing pronounced p-type behavior when subjected to excitation; the drain current experiences a substantial increase, signifying the generation of photocurrent in the HgTe CQDs device. The photoresponsivity of the device was maximum in the spectral range of 1600-2000 nm at 0.7 A/W under a –12 V gate and a –5 V drain condition [26].

Similarly, Zhang et al. [31] successfully developed high-performance deep ultraviolet (DUV) photodetectors utilizing graphene quantum dots (GQDs) through a simple solution process. The substantial band gap of GQDs allows operation in the DUV wavelength range, with responsivity increasing as the light wavelength decreases. Employing an asymmetric electrode structure, particularly Ag-Au electrodes, significantly enhances device sensitivity by reducing carrier transportation barriers compared to symmetric Au-Au or Ag-Ag electrodes [31]. Konstantatos et al. [32] present a report detailing the fabrication of solution-processed infrared photodetectors that surpass the normalized detectivity (D*) of the most advanced epitaxially grown devices operating at room temperature. These devices were created in a single solution-processing step by applying an unpatterned layer of PbS CQDs over a pre-fabricated planar electrode array. Notably, the devices exhibited substantial photoconductive gains, achieving responsivities exceeding 10^3 A/W. The most optimal devices demonstrated a remarkable normalized detectivity (D*) of 1.8×10^{13} Jones at 1.3 μm under room temperature conditions [32]. Recently, significant efforts have been devoted to enhancing the characteristics of CQDs devices. Notably, the device's responsibility has witnessed a substantial improvement, progressing from 10^{-3} to 10^3 A/W [33]. A detailed comparison table of CQD devices is provided below.

1.10 Photodetectors Based on 1D Nanomaterials

The major advantages of one-dimensional materials, for example, nanowires, tubes, and fibers lie in their extraordinary lengths, flexibility, and structures that enable physical manipulation of various shapes according to design requirements. Additionally, 1D structures exhibit unique quantum effects, leading to novel electronic properties [34].

For instance, Zhang et al. [35] conducted an investigation into the chiral-dependent optical properties of single-wall carbon nanotubes, with a focus on exploring their potential applications in wavelength-selective photodetectors, as shown in Figure 4a. The device fabrication process involved initially depositing a uniform SWCNTs network onto a poly-L-lysine functionalized SiO_2 surface, serving as the channel semiconductor. Following this, Ti/Pd source/drain electrodes were patterned. Subsequently, the thin film covering the entire substrate of SWCNTs underwent patterning using oxygen plasma to define the active channel region [35]. The AFM image reveals a uniform and dense distribution of SWCNTs, ensuring the proper electrical performance devices [35]. The I-V characteristics of the device were measured under bias voltage and gate voltage to ensure proper contact. Additionally, the switching characteristics of the device were investigated under varying excitation power at 650 nm, as illustrated in Figure 4c. It corresponds to increased photo response at higher laser intensities, which is well-known and documented [35]. The response time is estimated to be less than 5 ms at a gate voltage (V_G) of –5 V and between 9 to 11 ms at 7.5 V. These values fall within the typical range for bolometric effects [35]. Another key advantage of 1D nanostructures lies in their effectiveness in light trapping. This is achieved through mechanisms such as scattering, which increases the path length of light and generates strong electromagnetic fields within the active layer [82]. Additionally, 1D nanostructures facilitate multiple reflections and absorptions, enhancing their role in optimizing light absorption for various applications [38–40]. Vidur et al. [28] proposed a radial *p–n* junction using InP nanowire array photodetectors employing carrier-selective contact. The device, featuring *p*-type InP nanowire arrays with an *n*-type carrier-selective contact (ZnO), showcases the capability to detect single photons at room temperature without the need for external bias [41]. This remarkable performance is attributed to its high broadband absorption, exceptionally low dark current, and efficient charge carrier separation facilitated by a robust internal built-in electrical field exceeding 10^5 V cm^{-1} at zero bias [41]. 1D materials, particularly Carbon Nanotubes (CNTs), offer versatility for applications in flexible electronics. Kim et al. introduced a wearable UV sensor [42]. In this innovative design, a UV-sensitive SWCNTs ink was applied to coat a flexible thread substrate, as shown in Figure 4d [42]. By employing this process, the resulting thread not only attains flexibility but also experiences a substantial increase in sensitivity to ultraviolet (UV) radiation. Furthermore, silver paste is applied, acting as both the source and drain electrodes in the device [42]. Figure 4e (left) displays a scanning electron

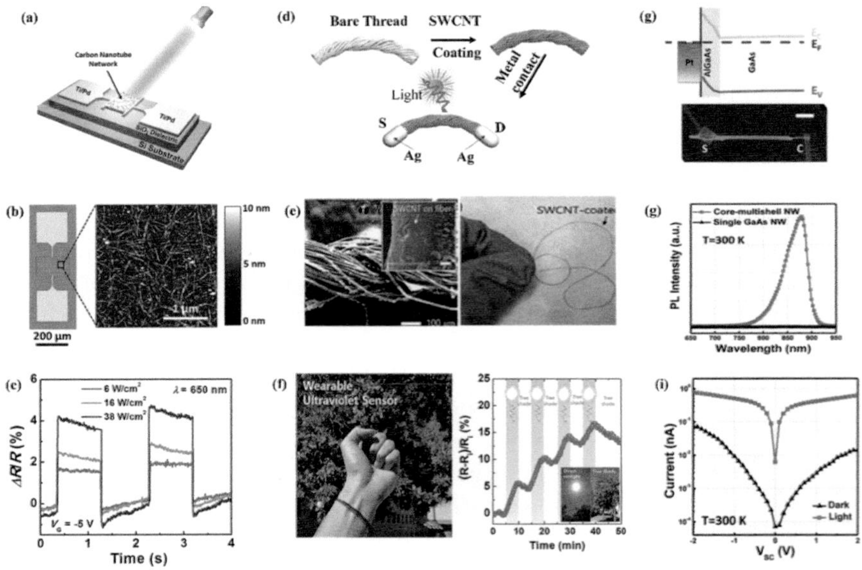

Figure 4. (a) Schematic Illustration of the single-wall carbon nanotubes (SWCNTs) based phototransistor, where Ti/Pd is a source and drain electrode. (b) Optical microscope image of the transistor with dimensions L = 100 µm and W = 200 µm. Inset: An AFM image displaying the network of single chirality nanotubes within the channel. (c) Dynamic photoresponse of the photodetector under different laser intensities of 6, 16, and ~38 W/cm^2 measured at a V_G of –5 V. (Figure a-c reprinted with permission. Copyright © 2015, Zhang et al. Springer Nature [35]). (d) Outline the step-by-step schematic illustration of the process to fabricate a cotton thread coated with single-walled carbon nanotubes (SWCNTs) based flexible photosensor. (e) Scanning electron microscopy (SEM) image of the cotton thread coated with SWCNTs. A high-resolution image and an optical micrograph of the SWCNTs coated thread are featured in the center. The left side displays an optical image of the light sensor, demonstrating its practical application as an ultraviolet light sensor. On the right side, the normalized resistance of the wearable UV sensor is depicted, measured alternately under direct sunlight and in tree shade. The inset features photographs illustrating the different UV measurement conditions. (Figures d-e were reprinted with permission from Kim et al. [36] Copyright © 2018, American Chemical Society.) (g) Illustration of the schematic band diagram depicting the zero-biased metal-semiconductor contact (top) and the SEM Image (bottom) of metal-semiconductor-metal (MSM) nanowire photodetector with selective contacts to the core (C) and the shell (S). The scale bar represents 2 µm. (h) A comparative analysis of micro-photoluminescence emission between a single GaAs nanowire and a core-multishell GaAs/high-T GaAs/AlGaAs nanowire at room temperature. (i) I-V characteristics of the photodetector depicted under dark conditions and with white light illumination. (Figures g-i reprinted with permission Zhang et al. [37] Copyright © 2017, American Chemical Society.)

microscopy (SEM) image of the same photograph (right) of the thread coated with SWCNTs [42]. Finally, the sensor device was tested under direct sunlight, confirming its practical usability for a wearable and flexible UV sensor system

[42]. The versatility of 1D materials allows for the engineering of core-shell devices, unlocking enhanced light-matter interaction and confinement, thereby broadening their applications in optoelectronics [43, 44]. A notable contribution in this realm comes from Dai et al. [44] where they have proposed a core-shell GaAs/AlGaAs nanowire photodetector designed for room-temperature operation. The device incorporates a type-I heterostructure within a metal-semiconductor-metal (MSM) radial architecture. The synergistic effect of built-in electric fields at the semiconductor heterointerface and the metal/semiconductor Schottky contact promotes efficient photogenerated charge separation, ultimately leading to augmented photosensitivity. The schematic illustrating the band alignment and device structure is presented in Figure 4g [44]. Before getting into the details of the device, a thorough investigation was carried out to explore the optical properties of both GaAs/AlGaAs and GaAs.Intriguingly, the photoluminescence of the core-shell GaAs/AlGaAs revealed a highly intense peak at 878 nm (~1.412 eV, represented by red squares). This observation underscores the pivotal role played by the AlGaAs shell in surface state passivation and carrier confinement, as depicted in Figure 3b. Indeed, in the case of pure GaAs, no emission peak is observed (Figure 4g), highlighting the enhanced and fascinating properties achieved through material engineering compared to individual components [44]. Finally, the I-V characteristics of the device were measured, revealing that the nanowire photodetector exhibits a photocurrent to dark current ratio of 145 at V_{SC} = 1 V. This ratio surpasses values reported for previously developed GaAs nanowire photodetectors [44]. The examples presented above prominently highlight the array of captivating features intrinsic to 1D materials, underscoring their potential applications across a wide range of devices.

1.11 Photodetectors Based on 2D Nanomaterials

2D materials have garnered significant attention from the scientific community in recent years thanks to their distinctive properties at the nanometer scale. Referred to as layered materials, they encompass various forms such as nanoribbons, nanoplates, nanowalls, and nanoflakes [45]. Three primary members of layered 2D materials are graphene, 2D chalcogenides, and 2D oxides [46,47]. These materials, characterized by their layered structure, offer the potential to be cleaved into individual freestanding atomic layers [47]. In Figure 5a, the straightforward architecture of a graphene-based photodetector is depicted. The left schematic illustrates the graphene/SiO_2/Si configuration, while the right showcases the graphene/Si. At the top, a gold electrode functions as both the source and drain. Initially, CVD graphene was transferred using a polymer-assisted method onto SiO_2/Si and Si substrates. Subsequently, photolithography was employed as a mask to define the positions of electrodes, followed by the deposition of Cr/Au films (5 nm/50 nm) on silicon oxide and Si as electrodes through the thermal evaporation deposition technique [48].

The photoresponse of both devices studied under the wavelength of 1550 nm, depicted as a function of time in Figure 5b, reveals a notable distinction.

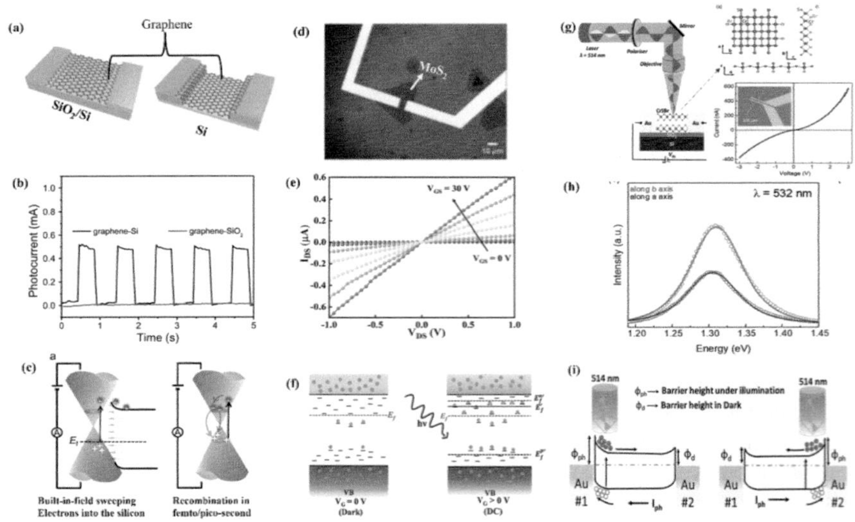

Figure 5. (a) The schematic illustration of the pristine graphene/SiO_2/Si and Graphene/Si-based device. (b) The photoresponse under laser pulses with a wavelength of 1550 nm is illustrated for both devices. (c), The impact of the built-in field at the graphene-silicon interface on photoinduced carriers in graphene is depicted (left), contrasting with the scenario of the device without a built-in field (right). (**F**igures a-c reprinted with permission from Chen et al. [48] Copyright © 2017, American Chemical Society.) (d) The optical microscopy image of a fabricated MoS_2 device reveals clear visibility of the source and drain electrodes on top of MoS_2 flakes. The scale bar is 10 μm. (e) The output characteristics of the MoS_2 transistor under dark conditions, varying V_{GS} from 0 to 30 V. (f) The schematic illustration of MoS_2 phototransistor under zero gate bias in dark conditions (Figure f, left); controlled trapping of photogenerated and intrinsic charge carriers in DC (Figure f, right). (Figures c-f reprinted with permission from Sahoo et al. [49] Copyright © 2023, American Chemical Society.) (g) The schematic illustration of the polarization measurement setup is presented on the left, where the excitation laser is linearly polarized by a half-wave plate before entering the confocal microscope. On the right, the crystal structure of CrSBr and an optical image of the device are shown. (h) The photoluminescence (PL) spectra of multilayer CrSBr under 532 nm excitation at different polarization axis. (i) The schematic illustration depicts the charge transfer and device mechanism under both dark and light conditions. (Figures g-i reprinted with permission from Panda et al. [50] Copyright © 2023, American Chemical Society.)

The pristine graphene device, as depicted by the red curve in Figure 5b, demonstrates almost no photocurrent. This well-known behavior is attributed to the natural zero band gap and rapid recombination of photo-induced carriers in graphene, occurring within femto/pico-seconds. The challenge lies in capturing the photocurrent signal before recombination. Consequently, the pristine graphene-based device exhibits poor responsivity [11,48]. In contrast, the graphene-silicon (Figure 5b, black curve) device demonstrates a clear and significant photocurrent. This distinction is attributed to the effective separation of photoinduced carriers by the built-in field at the interface of graphene-silicon. The observed phenomenon can be elucidated through the energy band diagram. In the presence of 1550 nm light illumination, photo-induced carriers are generated in graphene (Figure 5c left). Subsequently, electrons are trapped in silicon, as depicted in the schematic diagram. The resulting vacant holes in graphene contribute to a net increase in hole concentration, thereby generating the photocurrent. In contrast to pristine graphene (Figure 5c, right) photodetectors, which exhibit a small junction region near the electrodes, the significant built-in field along the entire graphene-silicon interface facilitates the effective separation of photoinduced carriers. This reduction in the recombination rate of spatially isolated photocarriers ultimately leads to a higher photocurrent [48]. Graphene, renowned for its remarkable properties, has garnered significant attention. However, its absence of an electronic bandgap has prompted the exploration of 2D materials with semiconducting characteristics. Transition metal dichalcogenides (TMDCs), belonging to the MX_2 semiconductor family, where M represents a transition metal atom (e.g., Mo or W) and X denotes a chalcogen atom (e.g., S, Se, or Te), emerge as a promising alternative. MoS_2, owing to its robustness, stands out as the most extensively studied material within this family. TMDCs present a unique combination of atomic-scale thickness, a direct bandgap, robust spin-orbit coupling, and favorable electronic and mechanical properties. These attributes render them compelling for fundamental studies and applications across diverse fields, including high-end electronics, spintronics, optoelectronics, energy harvesting, and flexible electronics [51]. The first MoS_2 phototransistor was reported by Sanchez et al. [52]. The monolayer MoS_2 photodetector exhibited a remarkable photoresponsivity of 880 A W^{-1}, marking a 100,000-fold improvement compared to previous reports on monolayer MoS_2 phototransistors [53]. This unprecedented enhancement is attributed to improved mobility, enhanced contact quality, and precise positioning techniques. Leveraging its direct bandgap, our ultrasensitive MoS_2 photodetectors showcase a photoresponsivity that surpasses the first graphene photodetectors by a factor of 10^6 (0.5 mAW^{-1}). The direct bandgap in monolayer MoS_2, as opposed to the indirect bandgap in bulk MoS_2, further contributes to a 9,000-fold higher photoresponsivity in monolayer MoS_2 compared to multilayer devices [52]. The response time of the devices was measured in seconds, with a rise time of 4 seconds and a fall time of 9 seconds [52]. The spectral response of the MoS_2 device spans from 400 to 680 nm, aligning with the absorption characteristics of MoS_2 [52]. A significant improvement has been achieved in pure MoS_2-based photodetectors [54]. Recently, Sahoo et al. [49] have reported a facile and effective approach to control carrier trapping, resulting in a highly responsive and low-power photodetection utilizing monolayer MoS_2 as the photoactive material. The optical image of the device is depicted in Figure 5d, showcasing the source and drain electrodes deposited on MoS_2 flakes through the nanofabrication technique [49]. The electrical characterization of the MoS_2 device is conducted at room temperature under high vacuum conditions (~10^{-5} mbar) in the dark. Figure 5e illustrates the linear output characteristics, depicting the source-to-drain current

(I_{DS}) with respect to the source-to-drain voltage (V_{DS}). This behavior describes the ohmic contact in the metal/MoS_2 junction [49, 53]. By implementing gate bias as a stream of pulses, the photogating and photoconductive effects are simultaneously enhanced, offering an opportunity to achieve high responsivity in optoelectronic devices. As a consequence, this approach yields a substantial photoresponsivity of approximately 4.2×10^3 A/W and a high photogain of around 11.3×10^3 with positive gate bias stress even under low-power illumination. Moreover, the gate-tunable photoresponse is demonstrated by achieving a photoresponsivity and photogain of approximately 0.7×10^3 A/W and 1.92×10^3, respectively, with negative gate bias stress [49]. The schematic illustration of the device mechanism is shown in Figure 5f. Recently discovered Janus 2D materials offer an enhanced capability to introduce an optically anisotropic response in monolayer 2D materials.[55] Specifically, the newly prepared CrSBr demonstrates an air-stable, optically anisotropic, semiconducting nature with a band gap of 1.5 [56]. Given the intriguing properties of CrSBr, we have designed an optical polarization-resolving photodetector (PD) employing multilayer semiconducting CrSBr single crystals, notable for their high structural anisotropy [44]. The schematic illustration of the device, along with the crystal structure of CrSBr, is presented in Figure 5g. This diagram showcases the mechanical transfer of multilayer CrSBr crystals onto a SiO_2/Si substrate with patterned electrodes facilitated by a transfer stage, as depicted in Figure 5g [50,57,58]. The I-V curve confirms a typical nonlinear diode-like characteristic, indicating the formation of Schottky junctions at the semiconducting CrSBr and Au interface (Figure 5g right).

The photoluminescence (PL) spectra of multilayer CrSBr, obtained under 532 nm excitation at room temperature, reveal a prominent emission centered at approximately 1.31 eV (Figure 5h). The observed emission is ascribed to band-edge emission rather than localized ligand-field luminescence. Additionally, the intensity of this emission exhibits variation with the crystal axis of CrSBr [50]. This self-powered photodetector operates through the formation of Schottky junctions at the Au-CrSBr interfaces, resulting in a position-sensitive and binary nature of the photocurrent [50]. The self-biased CrSBr PD exhibits a photoresponsivity of ~0.26 mA/W with a detectivity of 3.4×10^8 Jones at 514 nm excitation fluency (0.42 mW/cm^2) under ambient conditions. The optical polarization-induced photoresponse shows a large dichroic ratio of 3.4, with polarization set along the a and b-axes of single-crystalline CrSBr. Furthermore, the PD demonstrates excellent stability, retaining over 95% of the initial photoresponsivity in ambient conditions for more than five months without encapsulation. Therefore, in cutting-edge technology, we showcase CrSBr as a promising material for ultralow-powered optical polarization-resolving optoelectronic devices [50]. The charge transfer mechanism under different conditions is given in Figure 5i [50].

1.12 Photodetectors Based on Hybrid Nanomaterials

Indeed, the focus of the preceding discussion primarily centered on individual materials, particularly 0D, 1D, and 2D materials. It is noteworthy that individual materials inherently possess limited properties [59]. However, the creation of heterostructures or nanocomposites by combining materials of different dimensions results in innovative nanocomposites that often demonstrate enhanced physical and chemical properties compared to their individual counterparts. This enhancement broadens the scope of applications for such materials [59–64]. An illustrative example involves the combination of graphene quantum dots with 2D materials. GQDs have notable absorption in the UV region and less-than-optimal electrical properties. However, combining them with other 2D materials can significantly improve their overall properties. In this context, colloidal gold nanoparticle/graphene quantum dot (Au@GQD) nanohybrids under ambient conditions were produced using microplasmas, as depicted in Figure 6a. In this structure, the core consists of GQDs, while gold (Au) forms the shell. This design allows for an extended optical absorption range, covering the UV region and reaching the visible spectrum through plasmonic absorption [65]. These nanohybrids serve as photoabsorbers for broad-band PDs. The schematic illustration of the device is presented in Figure 6b, where graphene serves as the charge transport channel, facilitating the movement of charge carriers generated by Au@GQD upon exposure to light [65]. Notably, the Au@GQD-based visible light photodetector demonstrates remarkable photoresponsivity (approximately 10^3 A/W), ultrahigh detectivity (10^{13} Jones), and fast response times in the millisecond scale (65 ms rise time and 53 ms fall time). The synergistic effect contributing to these impressive characteristics is attributed to the strong fluorescence quenching in Au@GQD coupled with the two-dimensional graphene layer in the device. This research provides valuable insights into tailoring optical absorption in graphene quantum dots with plasmonic gold nanoparticles and the associated photophysics for achieving broad-band response in PD-related devices [65].

The comparative analysis between the GQD/graphene and Au@GQDs/graphene structures reveals the enhanced photocurrent in the presence of plasmonic gold, as depicted in Figure 6c.[65] Similarly, hybrid upconversion nanoparticle (UCNP)-graphene composites can serve for broad-band photodetection. The UCNPs, with a 980 nm multiphoton absorbing property, serve as the photoabsorber, while optimized graphene acts as an efficient charge transporter. The device benefits from critical engineering of the UCNPs, incorporating a silica (SiO_2) shell for optical coupling with graphene. This unique coupling results in both fluorescence quenching (using core UCNPs) and enhancement (using UCNP@SiO_2), significantly improving detector parameters. The photoresponsivity of the core-UCNP device, approximately ~1.52×10^4 AW^{-1}, is further enhanced to ~2.7×10^4 AW^{-1} (at 980 nm, power density of ~31.84 μW cm^{-2}, and under a 1.0 V bias) with the UCNP@SiO_2 device. The response time of the core-shell-based device is 2 to 8 seconds. This composite photodetector achieves the highest reported responsivity, gain, and detectivity in its class, capable of detecting signals from

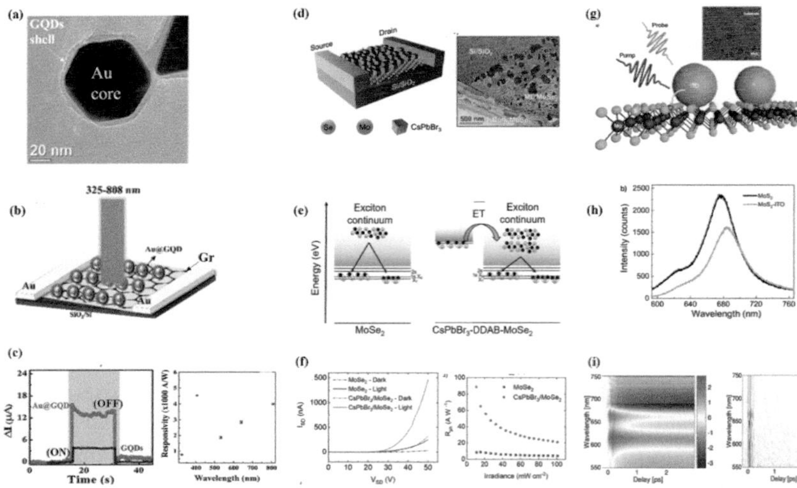

Figure 6. (a) The high-resolution TEM image displays the Au@GQD core/shell nanohybrid. (b) A schematic illustration depicts the Au@GQD/graphene hybrid devices on a SiO_2/Si substrate. (c) The comparison of photoresponse is presented between pure GQD/graphene and Au@GQD/graphene, with the responsivity shown as a function of excitation wavelength for the Au@GQD/graphene device on the right. (Figure 6a-c reprinted with permission from Thakur et al. [65]. Copyright © 2020, American Chemical Society.) (d) On the left, a schematic illustration depicts the $MoSe_2$/Perovskite nanocrystals-based device, while the corresponding SEM image of the hybrid device is shown on the right. (e) The schematic presents a one-particle energy level model of the heterostructure. On the left side, the one-particle depiction of monolayer $MoSe_2$) exhibits energy levels, including excitons and trions. The exciton continuum is positioned above. On the right side, an illustration showcases the energy transfer process (ET) from $CsPbBr_3$ nanocrystals to $MoSe_2$, resulting in enhanced carriers within the exciton continuum and trion formation. Grey energy levels representing DDAB ligands are depicted between the two materials. (f) The I-V characteristics of the hybrid device are depicted under both dark and light conditions on the left side, while the right side illustrates the photoresponsivity as a function of power for $MoSe_2$ and $CrPbBr_3/MoSe_2$. (Figure d-f reprinted with permission © 2022. Asaithambi et al. Advanced Optical Materials published by Wiley-VCH GmbH.) (g) The schematic illustration of the ITO NCs/MoS_2-based hybrid device and the corresponding SEM image of the heterostructure (inset) are presented. (h) Photoluminescence (PL) spectra of pure MoS_2 and ITO NCs/MoS_2 hybrid structure.(i) The differential transmissivity map of the heterostructure excited above the gap at 500 nm (left) and MoS_2 excited at 1750 nm (right) is depicted. (Figure 6g-i reprinted with permission from Guizzardi et al. [66] Copyright © 2022, American Chemical Society.)

common hand-held appliances [11]. To address the relatively slow response in the core-shell UCNPs graphene, further improvements were implemented by coupling gold nanorods (AuNRs) with UCNP@SiO_2 to enhance light-matter interaction [61]. The AuNR@UCNP@SiO_2 hybrid PD exhibited broadband photoresponse across the 455–980 nm range, boasting a photoresponsivity of approximately 5000

A/W. Furthermore, under 980 nm illumination, the response times ranged from 80 to 200 ms (1000 time faster than core-shell based device) [61]. A plethora of literature and publications is available on hybrid devices [60,67–71]. The hybrid structure is not limited to graphene; it can also be seamlessly integrated with TMDCs [60]. For example, the successful integration of $MoSe_2$ and perovskite ($CsPbBr_3$) nanocrystals [72]. The schematic illustration of the integrated device is shown in Figure 6d. The interaction between $CsPbBr_3$ nanocrystals and $MoSe_2$ monolayers results in an energy transfer efficiency exceeding 86%. This efficiency is determined from the quenching and decay dynamics observed in the emission of $CsPbBr_3$ nanocrystals. Notably, the increase in emission from the $MoSe_2$ monolayer in the heterostructure accounts for only 33% of the transferred energy [72]. The schematic illustration of the energy transfer process between $CsPbBr_3$ and $MoSe_2$ is depicted in Figure 6e. The efficacy of the heterostructure is further demonstrated through enhanced photocurrents (Figure 6f) compared to the individual material unit. These findings present a promising approach to address the high exciton binding energy characteristic of TMDCs, potentially impacting optoelectronic processes reliant on efficient exciton dissociation [72]. Indeed, several reports indicate that integrating TMDCs with plasmonic nanocrystals enhances the light-matter interaction, leading to increased photoresponse and improved response times in the devices [46,54,73–77]. The recent demonstration of integrating TMDCs with infrared plasmonic nanocrystals (NCs) signifies a promising advancement. As prominent example, Indium Tin Oxide (ITO) NCs hold substantial potential across various applications, particularly in fields such as energy storage devices and infrared photodetectors [78–81]. In recent developments, significant progress has been made in the integration of heavy metal-free infrared plasmonic ITO NCs with a MoS_2 monolayer. The heterostructure is created through a straightforward process involving the spin coating of ITO NCs onto chemical vapor deposition (CVD) grown MoS_2. The SEM image of the resulting heterostructure device is depicted in Figure 6g, with the inset providing a closer view [66]. The interaction of ITO NCs with MoS_2 is investigated with different techniques. The Photoluminescence (PL) spectra of both pure MoS_2 and ITO NCs/MoS_2 were measured, revealing a distinctive red-shift and broadening of the spectrum (Figure 6h). This phenomenon can be attributed to the additional trion PL resulting from doping the 2D material with ITO NCs [66]. The overall absorption of the heterostructure is also increased compare to pristine MoS_2. The investigation into charge transfer involved the utilization of ultrafast differential transmission, with sample excitation specifically set at 1750 nm (Figure 6i). This wavelength corresponds to the intense localized plasmon surface resonance of indium tin oxide nanocrystals (ITO NCs), exploiting the fact that monolayer molybdenum disulfide (MoS_2) does not absorb light at this particular wavelength (Figure 6i) [66]. The outcomes of this approach revealed the emergence of excitonic features of MoS_2 within the visible range, closely aligned with the MoS_2 exciton. This intriguing phenomenon has been attributed to a charge transfer mechanism occurring between the ITO NCS and monolayer MoS_2 upon plasmon excitation. These findings mark an initial step in exploring the utilization of plasmonic materials for enhanced photoconversion applications in the field of near-infrared photonics [66].

2. Conclusion and Future Outlook

In conclusion, the integration of low-dimensional nanomaterials into hybrid nanostructures presents a promising avenue for unlocking a broader range of enhanced material properties with diverse applications. Throughout this chapter, we have comprehensively explored nanomaterials, categorizing them into various types and examining different hybrid structures, particularly focusing on their applications in photodetectors.

2.1 Future Outlook

Looking towards the future, the integration of 0D and 2D materials holds significant potential for widespread applications, particularly in the telecommunications wavelength range. Stability emerges as a pivotal concern, notably in comparison to presently utilized materials such as silicon (Si). The infrared plasmonic nanocrystals, free of heavy metals, exhibit substantial potential in energy storage and photodetector applications. However, achieving smooth integration poses a critical challenge, and this hurdle can be addressed through effective interfacial engineering.

References

1. R.P. Feynman, There's plenty of room at the bottom, Eng. Sci., 23 (1959).
2. K.E. Geckeler, H. Nishide, Advanced Nanomaterials. John Wiley & Sons (2009).
3. Y. Gogotsi, Nanomaterials Handbook. CRC Press (2006).
4. M.C. Roco, Nanoparticles and nanotechnology research, J. Nanopart. Res., 1 (1999) 1.
5. D. Loss, Quantum phenomena in nanotechnology, Nanotechnology, 20 (2009) 430205.
6. C. Buzea, I. Pacheco, Nanomaterials and their classification, EMR/ESR/EPR Spectroscopy for Characterization of Nanomaterials (2017) 3-45.
7. T. Edvinsson, Optical quantum confinement and photocatalytic properties in two-, one- and zero-dimensional nanostructures, R. Soc. Open Sci., 5 (2018) 180387.
8. R.A. Yotter, D.M. Wilson, A review of photodetectors for sensing light-emitting reporters in biological systems, IEEE Sens J., 3 (2003) 288-303.
9. K. Gajanan, S.N. Tijare, Applications of nanomaterials, Mater. Today Proc., 5 (2018) 1093-1096.
10. B.E.A. Saleh, M.C. Teich, Fundamentals of Photonics. John Wiley & Sons (2019).
11. M.K. Thakur, A. Gupta, M.Y. Fakhri, R.S. Chen, C.T. Wu, K.H. Lin, S. Chattopadhyay, Optically coupled engineered upconversion nanoparticles and graphene for a high

responsivity broadband photodetector, Nanoscale, 11 (2019) 9716-9725. https://doi.org/10.1039/C8NR10280E.

12. D.N. Congreve, J. Lee, N.J. Thompson, E. Hontz, S.R. Yost, P.D. Reusswig, M.E. Bahlke, S. Reineke, T. Van Voorhis, M.A. Baldo, External quantum efficiency above 100% in a singlet-exciton-fission–based organic photovoltaic cell, Science (1979), 340 (2013) 334-337.
13. L. Brus, Electronic wave functions in semiconductor clusters: Experiment and theory, J. Phys. Chem., 90 (1986) 2555-2560. https://doi.org/10.1021/j100403a003.
14. C.B. Murray, D.J. Norris, M.G. Bawendi, Synthesis and characterization of nearly monodisperse CdE (E = sulfur, selenium, tellurium) semiconductor nanocrystallites, J. Am. Chem. Soc., 115 (1993) 8706-8715. https://doi.org/10.1021/ja00072a025.
15. M.M. Doeff, 2019 Nobel Prize in Chemistry. Nobel Prize (2019) 8.
16. D.V. Talapin, C.B. Murray, PbSe nanocrystal solids for n- and p-channel thin film field-effect transistors, Science (1979), 310 (2005) 86-89. https://doi.org/10.1126/science.1116703.
17. D. Yu, C. Wang, B.L. Wehrenberg, P. Guyot-Sionnest, Variable range hopping conduction in semiconductor nanocrystal solids, Phys. Rev. Lett., 92 (2004) 216802.
18. A. Pandey, P. Guyot-Sionnest, Slow electron cooling in colloidal quantum dots, Science (1979), 322 (2008) 929-932.
19. R.J. Ellingson, M.C. Beard, J.C. Johnson, P. Yu, O.I. Micic, A.J. Nozik, A. Shabaev, A.L. Efros, Highly efficient multiple exciton generation in colloidal PbSe and PbS quantum dots, Nano Lett., 5 (2005) 865-871. https://doi.org/10.1021/nl0502672.
20. V.I. Klimov, A.A. Mikhailovsky, S. Xu, A. Malko, J.A. Hollingsworth, C.A. Leatherdale, H.-J. Eisler, M.G. Bawendi, Optical gain and stimulated emission in nanocrystal quantum dots, Science (1979), 290 (2000) 314-317.
21. V.I. Klimov, A.A. Mikhailovsky, D.W. McBranch, C.A. Leatherdale, M.G. Bawendi, Quantization of multiparticle Auger rates in semiconductor quantum dots, Science (1979), 287 (2000) 1011-1013.
22. V.I. Klimov, Optical nonlinearities and ultrafast carrier dynamics in semiconductor nanocrystals, J, Phys, Chem. B, 104 (2000) 6112-6123.
23. R.D. Schaller, V.I. Klimov, High efficiency carrier multiplication in PbSe nanocrystals: Implications for solar energy conversion, Phys. Rev. Lett., 92 (2004) 186601.
24. G. Konstantatos, E.H. Sargent, Colloidal quantum dot photodetectors, Infrared Phys. Technol. 54 (2011) 278-282. https://doi.org/https://doi.org/10.1016/j.infrared.2010.12.029.
25. J. Chen, K. Žídek, P. Chábera, D. Liu, P. Cheng, L. Nuuttila, M.J. Al-Marri, H. Lehtivuori, M.E. Messing, K. Han, K. Zheng, T. Pullerits, Size- and wavelength-dependent two-photon absorption cross-section of $CsPbBr_3$ perovskite quantum dots, J. Phys. Chem. Lett., 8 (2017) 2316-2321. https://doi.org/10.1021/acs.jpclett.7b00613.
26. M. Chen, H. Lu, N.M. Abdelazim, Y. Zhu, Z. Wang, W. Ren, S.V. Kershaw, A.L. Rogach, N. Zhao, Mercury telluride quantum dot based phototransistor enabling high-sensitivity room-temperature photodetection at 2000 nm, ACS Nano, 11 (2017) 5614-5622. https://doi.org/10.1021/acsnano.7b00972.
27. Z. Hens, I. Moreels, Light absorption by colloidal semiconductor quantum dots, J. Mater. Chem., 22 (2012) 10406-10415. https://doi.org/10.1039/C2JM30760J.
28. Y. Pu, F. Cai, D. Wang, J.-X. Wan, J.-F. Chen, Colloidal synthesis of semiconductor quantum dots toward large-scale production: A review, Ind. Eng. Chem. Res., 57(6) (2018) 1790-1802.

29. A.P. Litvin, I.V. Martynenko, F. Purcell-Milton, A.V. Baranov, A.V. Fedorov, Y.K. Gun'ko, Colloidal quantum dots for optoelectronics, J. Mater. Chem. A Mater., 5 (2017) 13252-13275. https://doi.org/10.1039/C7TA02076G.
30. C.R. Kagan, L.C. Bassett, C.B. Murray, S.M. Thompson, Colloidal quantum dots as platforms for quantum information science, Chem. Rev., 121 (2021) 3186-3233. https://doi.org/10.1021/acs.chemrev.0c00831.
31. Q. Zhang, J. Jie, S. Diao, Z. Shao, Q. Zhang, L. Wang, W. Deng, W. Hu, H. Xia, X. Yuan, S.-T. Lee, Solution-processed graphene quantum dot deep-UV photodetectors, ACS Nano, 9 (2015) 1561-1570. https://doi.org/10.1021/acsnano.5b00437.
32. G. Konstantatos, I. Howard, A. Fischer, S. Hoogland, J. Clifford, E. Klem, L. Levina, E.H. Sargent, Ultrasensitive solution-cast quantum dot photodetectors, Nature, 442 (2006) 180-183. https://doi.org/10.1038/nature04855.
33. S. Miao, Y. Cho, Toward green optoelectronics: Environmental-friendly colloidal quantum dots photodetectors, Front Energy Res., 9 (2021). https://www.frontiersin.org/articles/10.3389/fenrg.2021.666534.
34. J. Hu, T.W. Odom, C.M. Lieber, Chemistry and physics in one dimension: synthesis and properties of nanowires and nanotubes, Acc Chem. Res., 32 (1999) 435-445. https://doi.org/10.1021/ar9700365.
35. S. Zhang, L. Cai, T. Wang, R. Shi, J. Miao, L. Wei, Y. Chen, N. Sepúlveda, C. Wang, Bolometric-effect-based wavelength-selective photodetectors using sorted single chirality carbon nanotubes, Sci. Rep., 5 (2015) 17883. https://doi.org/10.1038/srep17883.
36. S.J. Kim, D.-I. Moon, M.-L. Seol, B. Kim, J.-W. Han, M. Meyyappan, Wearable UV sensor based on carbon nanotube-coated cotton thread, ACS Appl. Mater. Interfaces, 10 (2018) 40198-40202. https://doi.org/10.1021/acsami.8b16153.
37. M. Abniel, K. Fontánez, J.C. Arango, D. Ortiz, J. De León, S. Pinilla, V. Nicolosi, F.I. Petrescu, C. Morant, F. Márquez, One-dimensional (1D) nanostructured materials for energy applications. Materials, 14(10) (2021) 2609.
38. H. Chen, L. Shao, Q. Li, J. Wang, Gold nanorods and their plasmonic properties, Chem. Soc. Rev., 42 (2013) 2679-2724.
39. F.-H. Kao, N. Akhtar, C.-C. Chen, H.Y. Chen, M.K. Thakur, Y.-Y. Chen, C.-L. Chen, S. Chattopadhyay, In vivo and in vitro demonstration of gold nanorod aided photothermal presoftening of B16F10 melanoma for efficient chemotherapy using doxorubicin loaded graphene oxide, ACS Appl. Bio. Mater., 2 (2019) 533-543. https://doi.org/10.1021/acsabm.8b00701.
40. C. Fei Guo, T. Sun, F. Cao, Q. Liu, Z. Ren, Metallic nanostructures for light trapping in energy-harvesting devices, Light Sci. Appl., 3 (2014) e161-e161. https://doi.org/10.1038/lsa.2014.42.
41. Y. Zhu, V. Raj, Z. Li, H.H. Tan, C. Jagadish, L. Fu, Self-powered InP nanowire photodetector for single-photon level detection at room temperature, Adv. Mater., 33 (2021) 2105729. https://doi.org/https://doi.org/10.1002/adma.202105729.
42. H.C. Ates, P.Q. Nguyen, L. Gonzalez-Macia, E. Morales-Narváez, F. Güder, J.J. Collins, C. Dincer, End-to-end design of wearable sensors, Nat. Rev. Mater., 7 (2022) 887-907.
43. C. Soci, A. Zhang, X.-Y. Bao, H. Kim, Y. Lo, D. Wang, Nanowire photodetectors, J. Nanosci. Nanotechnol., 10 (2010) 1430-1449.
44. X. Dai, S. Zhang, Z. Wang, G. Adamo, H. Liu, Y. Huang, C. Couteau, C. Soci, GaAs/AlGaAs nanowire photodetector, Nano Lett., 14 (2014) 2688-2693. https://doi.org/10.1021/nl5006004.

45. M. Zhou, X.W. (David) Lou, Y. Xie, Two-dimensional nanosheets for photoelectrochemical water splitting: Possibilities and opportunities, Nano Today, 8 (2013) 598-618. https://doi.org/https://doi.org/10.1016/j.nantod.2013.12.002.
46. X. Li, H. Zhu, Two-dimensional MoS_2: Properties, preparation, and applications, J. Materiomics, 1 (2015) 33-44. https://doi.org/https://doi.org/10.1016/j.jmat.2015.03.003.
47. K.S. Novoselov, A.K. Geim, S.V. Morozov, D. Jiang, Y. Zhang, S.V. Dubonos, I.V. Grigorieva, A.A. Firsov, Electric field effect in atomically thin carbon films, Science (1979), 306 (2004) 666-669. https://doi.org/10.1126/science.1102896.
48. Z. Chen, X. Li, J. Wang, L. Tao, M. Long, S.-J. Liang, L.K. Ang, C. Shu, H.K. Tsang, J.-B. Xu, Synergistic effects of plasmonics and electron trapping in graphene short-wave infrared photodetectors with ultrahigh responsivity, ACS Nano, 11 (2017) 430-437. https://doi.org/10.1021/acsnano.6b06172.
49. S. Sahoo, M.C. Sahu, S.K. Mallik, A.K. Jena, G.K. Pradhan, S. Sahoo, High responsivity in monolayer MoS_2 photodetector via controlled interfacial carrier trapping, ACS Appl. Electron Mater., 5 (2023) 1077-1087. https://doi.org/10.1021/acsaelm.2c01567.
50. J. Panda, S. Sahu, G. Haider, M.K. Thakur, K. Mosina, M. Velický, J. Vejpravova, Z. Sofer, M. Kalbáč, Polarization-resolved position-sensitive self-powered binary photodetection in multilayer Janus CrSBr, ACS Appl. Mater. Interfaces, 16 (2024) 1033-1043. https://doi.org/10.1021/acsami.3c13552.
51. S. Manzeli, D. Ovchinnikov, D. Pasquier, O.V. Yazyev, A. Kis, 2D transition metal dichalcogenides, Nat. Rev. Mater., 2 (2017) 17033. https://doi.org/10.1038/natrevmats.2017.33.
52. O. Lopez-Sanchez, D. Lembke, M. Kayci, A. Radenovic, A. Kis, Ultrasensitive photodetectors based on monolayer MoS_2, Nat. Nanotechnol., 8 (2013) 497-501. https://doi.org/10.1038/nnano.2013.100.
53. Z. Yin, H. Li, H. Li, L. Jiang, Y. Shi, Y. Sun, G. Lu, Q. Zhang, X. Chen, H. Zhang, Single-layer MoS_2 phototransistors, ACS Nano, 6 (2012) 74-80. https://doi.org/10.1021/nn2024557.
54. R. Wadhwa, A.V. Agrawal, M. Kumar, A strategic review of recent progress, prospects and challenges of MoS_2-based photodetectors, J. Phys. D Appl. Phys., 55 (2022) 063002. https://doi.org/10.1088/1361-6463/ac2d60.
55. L. Zhang, Z. Yang, T. Gong, R. Pan, H. Wang, Z. Guo, H. Zhang, X. Fu, Recent advances in emerging Janus two-dimensional materials: From fundamental physics to device applications, J. Mater. Chem. A Mater., 8 (2020) 8813-8830. https://doi.org/10.1039/D0TA01999B.
56. D.J. Rizzo, A.S. McLeod, C. Carnahan, E.J. Telford, A.H. Dismukes, R.A. Wiscons, Y. Dong, C. Nuckolls, C.R. Dean, A.N. Pasupathy, X. Roy, D. Xiao, D.N. Basov, Visualizing atomically layered magnetism in CrSBr, Adv. Mater., 34 (2022) 2201000. https://doi.org/https://doi.org/10.1002/adma.202201000.
57. V. Varade, G. Haider, L. Pirker, J. Panda, J. Sykora, O. Frank, M. Kalbac, J. Vejpravova, Sulfur isotope engineering of exciton and lattice dynamics in mathrm{MoS}_{2} monolayers, 2d Mater, 10 (2023) 025024. https://doi.org/10.1088/2053-1583/acc4d8.
58. M.K. Thakur, G. Haider, F.J. Sonia, J. Plšek, A. Rodriguez, V. Mishra, J. Panda, O. Gedeon, M. Mergl, O. Volochanskyi, V. Valeš, O. Frank, J. Vejpravova, M. Kalbáč, Isotope engineered fluorinated single and bilayer graphene: Insights into fluorination selectivity, stability, and defect passivation, Small, 19 (2023) 2205575. https://doi.org/https://doi.org/10.1002/smll.202205575.

59. M.K. Thakur, A. Gupta, S. Ghosh, S. Chattopadhyay, Graphene-conjugated upconversion nanoparticles as fluorescence-tuned photothermal nanoheaters for desalination, ACS Appl. Nano Mater., 2 (2019) 2250-2259. https://doi.org/10.1021/acsanm.9b00186.
60. A. Gupta, S. Ghosh, M.K. Thakur, J. Zhou, K. (Ken) Ostrikov, D. Jin, S. Chattopadhyay, Up-conversion hybrid nanomaterials for light- and heat-driven applications, Prog. Mater. Sci., 121 (2021) 100838. https://doi.org/https://doi.org/10.1016/j.pmatsci.2021.100838.
61. A. Gupta, M.K. Thakur, T.A. Effendi, R.-S. Chen, H.-Y. Cheng, K.-H. Lin, M. Bouras, D.S. Tomar, H.Y. Kuo, S. Chattopadhyay, Metallo-graphene enhanced upconversion luminescence for broadband photodetection under polychromatic illumination, Chemical Engineering Journal 420 (2021) 127608. https://doi.org/https://doi.org/10.1016/j.cej.2020.127608.
62. R. Parmar, S.J. Rezvani, D.B. de F. Neto, J.M. Rosolen, S. Kazim, S. Mattiello, P. Rajak, R. Ciancio, M.K. Thakur, M. Minicucci, M. Amati, L. Gregoratti, M. Kalbac, G. Haider, A. Di Cicco, R. Gunnella, Structural phase stability and homogeneity enhancement of electrochemically synthesized $Mn_2V_2O_7$ by nanocarbon networks, Carbon Trends, 9 (2022) 100218. https://doi.org/https://doi.org/10.1016/j.cartre.2022.100218.
63. S. Rathore, D.K. Patel, M.K. Thakur, G. Haider, M. Kalbac, M. Kruskopf, C.-I. Liu, A.F. Rigosi, R.E. Elmquist, C.-T. Liang, P.-D. Hong, Highly sensitive broadband binary photoresponse in gateless epitaxial graphene on 4H-SiC, Carbon N Y, 184 (2021) 72-81. https://doi.org/https://doi.org/10.1016/j.carbon.2021.07.098.
64. P. Kaushik, F.J. Sonia, G. Haider, M.K. Thakur, V. Valeš, J. Kong, M. Kalbáč, Toward graphene-enhanced spectroelectrochemical sensors, Adv. Mater. Interfaces, 9 (2022) 2200478. https://doi.org/https://doi.org/10.1002/admi.202200478.
65. M.K. Thakur, C.-Y. Fang, Y.-T. Yang, T.A. Effendi, P.K. Roy, R.-S. Chen, K.K. Ostrikov, W.-H. Chiang, S. Chattopadhyay, Microplasma-enabled graphene quantum dot-wrapped gold nanoparticles with synergistic enhancement for broad band photodetection, ACS Appl. Mater. Interfaces, 12 (2020) 28550-28560. https://doi.org/10.1021/acsami.0c06753.
66. M. Guizzardi, M. Ghini, A. Villa, L. Rebecchi, Q. Li, G. Mancini, F. Marangi, A.M. Ross, X. Zhu, I. Kriegel, F. Scotognella, Near-infrared plasmon-induced hot electron extraction evidence in an indium tin oxide nanoparticle/monolayer molybdenum disulfide heterostructure, J. Phys. Chem. Lett., 13 (2022) 9903-9909. https://doi.org/10.1021/acs.jpclett.2c02358.
67. Z. Luo, Y. Zhang, C. Zhang, H.T. Tan, Z. Li, A. Abutaha, X. Wu, Q. Xiong, K.A. Khor, K. Hippalgaonkar, Multifunctional 0D-2D Ni2P nanocrystals–black phosphorus heterostructure, Adv. Energy Mater., 7 (2017) 1601285.
68. A. Boulesbaa, K. Wang, M. Mahjouri-Samani, M. Tian, A.A. Puretzky, I. Ivanov, C.M. Rouleau, K. Xiao, B.G. Sumpter, D.B. Geohegan, Ultrafast charge transfer and hybrid exciton formation in 2D/0D heterostructures, J. Am. Chem. Soc., 138 (2016) 14713-14719.
69. S. Mukherjee, D. Bhattacharya, S. Patra, S. Paul, R.K. Mitra, P. Mahadevan, A.N. Pal, S.K. Ray, High-responsivity gate-tunable ultraviolet–visible broadband phototransistor based on graphene–WS2 mixed-dimensional (2D-0D) heterostructure, ACS Appl. Mater. Interfaces, 14 (2022) 5775-5784.

70. S. Sinha, H. Kim, A.W. Robertson, Preparation and application of 0D-2D nanomaterial hybrid heterostructures for energy applications, Mater. Today Adv., 12 (2021) 100169.
71. Z. Liang, M. Wang, X. Zhang, Z. Li, K. Du, J. Yang, S.-Y. Lei, G. Qiao, J.Z. Ou, G. Liu, A 2D-0D-2D sandwich heterostructure toward high-performance room-temperature gas sensing, ACS Nano (2024).
72. A. Asaithambi, N. Kazemi Tofighi, N. Curreli, M. De Franco, A. Patra, N. Petrini, D. Baranov, L. Manna, F. Di Stasio, I. Kriegel, Generation of free carriers in $MoSe_2$ monolayers via energy transfer from $CsPbBr_3$ Nanocrystals, Adv. Opt. Mater., 10 (2022) 2200638. https://doi.org/https://doi.org/10.1002/adom.202200638.
73. R. Wadhwa, A.V. Agrawal, M. Kumar, A strategic review of recent progress, prospects and challenges of MoS_2-based photodetectors, J. Phys. D Appl. Phys., 55 (2022) 063002. https://doi.org/10.1088/1361-6463/ac2d60.
74. V. Varade, G. Haider, A. Slobodeniuk, R. Korytar, T. Novotny, V. Holy, J. Miksatko, J. Plsek, J. Sykora, M. Basova, M. Zacek, M. Hof, M. Kalbac, J. Vejpravova, Chiral light emission from a hybrid magnetic molecule–monolayer transition metal dichalcogenide heterostructure, ACS Nano, 17 (2023) 2170-2181. https://doi.org/10.1021/acsnano.2c08320.
75. Y. Shi, J.-K. Huang, L. Jin, Y.-T. Hsu, S.F. Yu, L.-J. Li, H.Y. Yang, Selective decoration of Au nanoparticles on monolayer MoS_2 single crystals, Sci. Rep. 3 (2013) 1839. https://doi.org/10.1038/srep01839.
76. O. Lopez-Sanchez, D. Lembke, M. Kayci, A. Radenovic, A. Kis, Ultrasensitive photodetectors based on monolayer MoS_2, Nat. Nanotechnol., 8 (2013) 497-501. https://doi.org/10.1038/nnano.2013.100.
77. Y. Li, J.D. Cain, E.D. Hanson, A.A. Murthy, S. Hao, F. Shi, Q. Li, C. Wolverton, X. Chen, V.P. Dravid, Au@MoS2 core–shell heterostructures with strong light–matter interactions, Nano Lett., 16 (2016) 7696-7702. https://doi.org/10.1021/acs.nanolett.6b03764.
78. I. Kriegel, Physical phenomena behind photodoping in doped metal oxide nanocrystals, Proc. SPIE, 2022: p. PC1214201. https://doi.org/10.1117/12.2625984.
79. N. Petrini, M. Ghini, N. Curreli, I. Kriegel, Optical modeling of plasmonic nanoparticles with electronically depleted layers, J. Phys. Chem. C, 127 (2023) 1576-1587. https://doi.org/10.1021/acs.jpcc.2c05582.
80. A. Camellini, L. Rebecchi, A. Rubino, W. Niu, S.W. Kim, J. Ma, X. Feng, I. Kriegel, Improving the stability of photodoped metal oxide nanocrystals with electron donating graphene quantum dots, Nanoscale, 15 (2023) 17138-17146. https://doi.org/10.1039/D3NR03534D.
81. L. Rebecchi, A. Rubino, A. Camellini, I. Kriegel, Light-driven reversible charge transfers from ITO nanocrystals, Front. Chem., 11 (2023). https://www.frontiersin.org/articles/10.3389/fchem.2023.1288681.
82. X. Cui, Q. Ruan, X. Zhuo, X Xia. J. Hu, R. Fu, Y. Li, J. Wang, H. Xu, Photothermal nanomaterials: A powerful light-to-heat converter, Chemical Reviews, 123(11) (2023) 6891-6952.
83. B.B. Wang, R.W. Shao, X. Huang, K. Zheng, K. Zhu, Q.J. Cheng, Structure and photoluminescence properties of graphene nanoflakes grown on zinc oxide films by hot filament chemical vapor deposition, Diam. Relat. Mater., 64 (2016) 42-48.

CHAPTER
5

Nanostructured Metal Oxides for Energy Harvesting and Storage Devices

Jyoti Bala[1], Mohd. Shkir[2] and Ashwani Kumar[1*]

[1] Department of Physics, Goswami Ganesh Dutta Sanatan Dharma College, Chandigarh 160030, India

[2] Department of Physics, College of Science, King Khalid University, Abha 61413, Saudi Arabia

1. Introduction

In the contemporary global context, energy is a vital and complex factor that impacts social, economic, and environmental domains. Energy is essential to the expansion and development of the economy. Reliable and sufficient energy sources support greater output, the development of jobs, and general economic prosperity. Energy-intensive technologies also raise living standards generally in developed areas by promoting comfort, convenience, and technological progress. A reliable and strong energy source is essential for the development of technology, including digitalization, automation, and the expansion of the internet. Numerous cutting-edge technologies are changing sectors and requiring significant energy inputs, such as sophisticated manufacturing, renewable energy systems, and electric cars. An important factor to take into account is how energy production and consumption affect the environment. In order to prevent climate change, reduce air pollution, and conserve ecosystems, there is an urgent global need to shift to cleaner and more sustainable energy sources. In order to address environmental issues, renewable energy technology, energy efficiency measures, and sustainable practices are becoming more and more popular. The globe is moving away from conventional energy systems that rely on fossil fuels and towards greener, more sustainable options. Concerns about climate change, the depletion of limited resources, and the requirement for energy independence

*Corresponding author: ashw.chauhan@gmail.com

are what are driving this shift. An essential part of this worldwide change is the use of renewable energy sources, energy storage technology, and increased energy efficiency. In conclusion, energy is a fundamental component of modern society, impacting advancements in technology, economic growth, and general well-being.

The use of nanostructured metal oxides (MOs) in energy harvesting and storage devices has attracted a lot of interest recently. These materials exhibit unique properties at the nanoscale, which make them promising candidates for improving the performance of devices such as batteries, supercapacitors, fuel cells and solar cells [1–4]. The surface-to-volume ratio of nanostructured MOs is higher than that of their bulk equivalents. More active sites for reactions can occur due to the increased surface area, which is very beneficial for energy storage applications. Improved electrochemical characteristics, such as increased capacitance in supercapacitors and enhanced cycling stability in batteries, are frequently the consequence of nanostructuring MOs. The overall effectiveness and durability of energy storage systems are enhanced by these advancements. Controlled doping and alloying are made possible by nanostructuring, which enables changes to the MO's overall characteristics and electrical structure. This can be used to modify the material for use in energy devices in particular applications. These materials' nanoscale size frequently results in structures that are flexible and light. This helps to promote wearable electronics and flexible energy storage solutions, and it is especially beneficial for the development of portable and flexible energy storage systems [5]. These MO nanoparticles also improve thermoelectric performance for waste heat conversion and work well as agents for pollutant photocatalysis and environmental remediation [6, 7]. They also function as sensitive materials in environmental sensors that track pollutants in real time.

Numerous methods, such as sol-gel processes, hydrothermal synthesis, and template-assisted approaches [8–11], can be used to synthesize nanostructured MOs. Because of the material's adaptability during synthesis, researchers can customize its features for certain uses. Zinc oxide (ZnO), manganese oxide (MnO_2), iron oxide (Fe_2O_3), and titanium dioxide (TiO_2) are examples of transition MOs that are frequently researched as nanostructured MOs [12–17]. Even though nanostructured MOs have a lot of potential, issues including scalability, affordability, and long-term stability still need to be resolved before they can be widely used in commercial settings. In order to overcome these obstacles and realize the full potential of nanostructured MOs in energy devices, materials science and nanotechnology research is still underway.

2. Solar Cells

While linked devices can gather energy in a variety of ways, photovoltaic (PV) cell technology is expected to have the most uptakes in the upcoming years. Photogeneration of charge carriers in a light-absorbing substance and their separation into conductive materials to deliver power are the two primary processes of a solar cell. The majority of carriers are produced close to the surface. Electron–hole pairs are created when a light quantum strikes a semiconductor, greatly increasing the conductivity. They may arrive at the *p–n* junction, where the electric field divides the charges, if their recombination is stopped. These MOs are used in numerous solar cells, but in perovskite solar cells they are used widely and responsible for high efficiency of these cells [18,19]. The energy diagram for PSCs shown in Fig. 1, the working principle can be defined as follows, when a visible light was an incident; the photons were absorbed and generated the excitons (e^-, h^+ pairs).

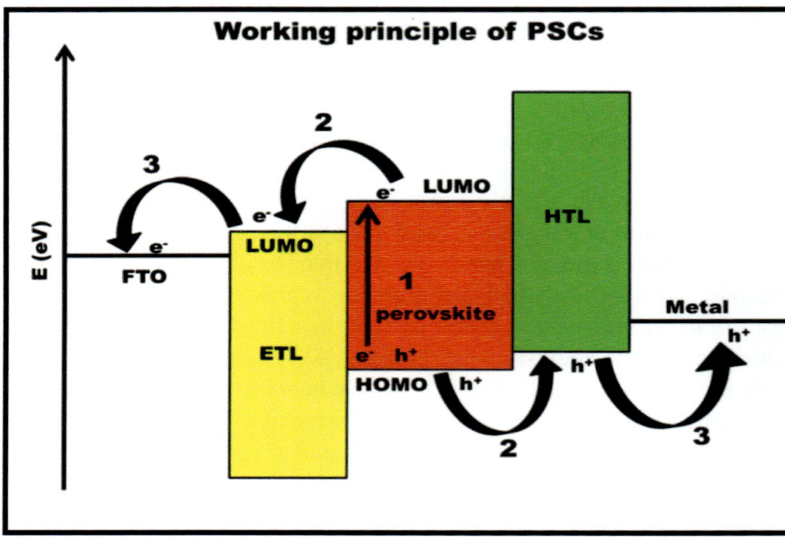

Figure 1. Schematic diagram for working principle of PSCs.

The photon-excited electrons have moved towards the conduction band (CB), and the holes are moved towards the valance band (VB). The CB of the perovskite is higher than that of the electron transport layer (ETL) (such as TiO_2, ZnO) [20], the photon-excited electron can diffuse into the BL. Simultaneously, the photon-excited hole is absorbed by the HTL.

MOs contribute significantly to the functioning and efficiency of solar cells in a variety of ways. The charge transport layer (CTL) is seeing a rise in the use of MOs due to their low cost of manufacturing, solution process ability, strong material stability, and adaptability in material change. The CTL is essential to the creation of perovskite solar cells (PSCs) because it makes it easier to recover charge carriers that are photo generated from the perovskite layer. Scalable fabrication is required for large-area PSCs in order to produce consistent and ideal CTLs. Even while PSCs have previously shown power conversion efficiencies

(PCE) above 26% in the lab, scaling up production is still quite difficult, which makes them less viable for commercial use. In actuality, the deployment of MOs depends on the PSCs' quick development of performance.

In solar cells, MOs are frequently employed as doping agents, blocking layers, transparent conducting oxides (TCOs), hole transporting materials (HTMs), and electron transporting materials (ETMs). Because MOs have more degrees of freedom, it is possible to modify optoelectronic and other properties more effectively. TCOs are substances that are electrically conductible and transparent to visible light. TCOs are mainly used to provide a conductive conduit for the electrical current generated while letting sunlight reach the active layers. One of the TCOs that is most frequently utilised is indium tin oxide (ITO). It is a mixture of tin oxide (SnO_2) and indium oxide (In_2O_3). Another TCO that is frequently utilized is fluorine-Doped tin oxide (FTO). Its foundation is tin oxide that has been doped with fluorine to increase conductivity. FTO is frequently utilized as an alternative since it is less expensive than ITO. Mechanical flexibility is a prerequisite for TCOs utilized in flexible solar cells. To create flexible TCOs that can continue to be transparent and conductive when bent, research is still being done. Zinc oxide (ZnO) exhibited broad band gap and *n*-type semiconductor materials. ZnO films demonstrate great carrier mobility, ease of crystallisation, and technical adaptability. ZnO films have a very smooth, compact, and uniform morphology, which makes them suitable for use as an electron transport layer in perovskite solar cells (PSCs), as noted by Sudheer and colleagues [21]. Aside from this, TiO_2 is currently the most often utilised ETL for producing PSCs [22]. In every PSC construction, it has typically been utilised to block holes and promote constructive electron transmission. When TiO_2 is utilised as an ETL layer in a device, there have been certain disadvantages. Due to the deterioration of the perovskite materials, interstitials such as titanium and oxygen vacancies may occur in this layer, which reduces the effectiveness of PSCs. In order to solve this issue, TiO_2 has been doped into aluminium (Al), increasing the composition's conductivity and improving device performance [23,24]. Better than TiO_2, the ZnO-Ar film mixture provides planar PSCs by acting as a hole blocker and electron extractor. The *n*-type semiconductors ZnO and Fe_3O_4 among these oxides have a broad bandgap of 2–3.5 eV, making them attractive ETLs for perovskited electrodes because of their inherent qualities, which include a high electron transport rate of 130–230 $cm^2\,V^{-1}s^{-1}$ and strong chemical stability. A recent work on spin coating large-area ETLs found that the SnO_2-TiO_2 hybrid ETL makes it easier to use this coating approach to create flexible, effective perovskite mini-modules. Numerous research groups have also looked into WO_x as an alternative to TiO_2 in PSCs. Despite having a lower band gap than TiO_2, it exhibits great stability and a higher electron mobility (10–20 $cm^2V^{-1}s^{-1}$) than TiO_2. Comparing the 2-D nanosheet-WO_3 film to nanorods or nanoparticles, more carrier transport was produced. Nb_2O_5 has been studied as an ETL and hole-blocking layer in PSCs due to its broad optical band gap (4.1 eV), strong transmission, greater carrier mobility, and appropriate conduction band energy level.

According to a number of experts, nickel oxide (NiO) was crucial as a material for hole transmission. Due to a number of special qualities, including cheap cost, little hysteresis, a suitable band gap, strong photo stability, good chemical stability [25], outstanding transmittance, low electrode polarisation, and a reduced processing temperature for the fabrication of solar cells. Furthermore, a method for copper doping and seeding-induced crystallization is presented. Using the seed-assisted induction method produces a compact NiOx layer with fewer defect sites by inducing a high grain size through controlled crystal development. In order to enhance hole transport and conductivity, a Cu-doping method is utilised to encourage the creation of shallow acceptor levels, which serve as defect states. Outstanding PCEs of 22.51% and 19.29% were obtained for small-area mini-modules (0.06 cm^2) and large-area mini-modules (active area of 28 cm^2), respectively, using an ideal Cu doping concentration.

The researchers came to the conclusion that by covering the transparent conductive oxide (TCO) layer with a dielectric coating (like Al_2O_3), the efficiency of the solar cell might be increased. High dielectric strength, strong stability, outstanding transparency, and resilience under extreme circumstances were all demonstrated by the Al_2O_3 [26]. Using the atomic layer deposition approach, Choi et al. [27] generated Al_2O_3, which could be employed as a protective layer for PSCs. In polymer solar cells, solution-processed tungsten oxide films have been used as the active anode buffer layer. Figure 2 gives details of an energy level diagram of different MOs used as ETL materials in PSCs. Besides this, light absorption in many solar cells produces electron-hole pairs in the semiconductor

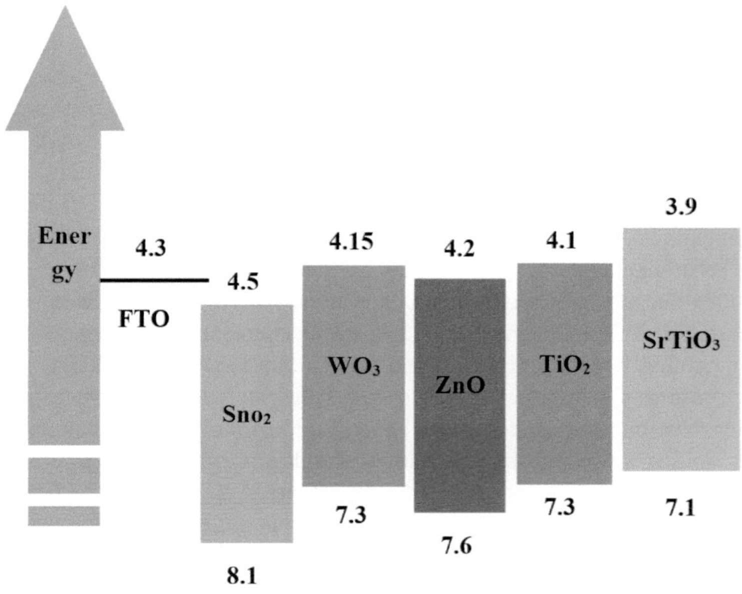

Figure 2. Energy diagrams of different MOs ETL layers used in PSCs.

material. The effective separation of these charge carriers is aided by MOs. For example, when photons enter the semiconductor layer, excited electrons migrate to the MO layer, leaving the semiconductor with positively charged holes. In order to produce an electric current, these charges must separate. MOs have the ability to possess semiconductor layer surface imperfections. By creating a thin layer at the interface between the semiconductor and the MO, passivation lowers the density of electronic faults. By reducing electron and hole recombination, this raises the solar cell's total efficiency. It is possible to engineer MOs to have particular optical characteristics. Certain light wavelengths can pass through them, enabling sunlight to reach the semiconductor layer for absorption. In order to retain good electron transport characteristics and allow photon absorption in the semiconductor, transparency is essential. In tandem solar cells, where many layers of various semiconductor materials are layered to capture a wider range of the solar spectrum, MOs are also used. Between these layers, MOs aid in preserving electrical continuity. To summarize, the presence of MOs in solar cells aids in several processes such as electron transport, charge separation, defect passivation, optical qualities, stability, and durability. They do a variety of tasks that make them essential to the design and development of different kinds of solar cells. In conclusion, PSCs longevity is contingent upon the integrity of the interface (top electrode or bottom electrode/adjacent layer, absorber/ETL or HTL). To increase the stability of PSCs, special attention must be paid to stable electrodes, ETL, HTL, absorbers, and the deposition of these layers. This idea also holds true for other third-generation solar cells. Figure 3 gives details with an energy level diagram of different MOs used as HTL materials in PSCs.

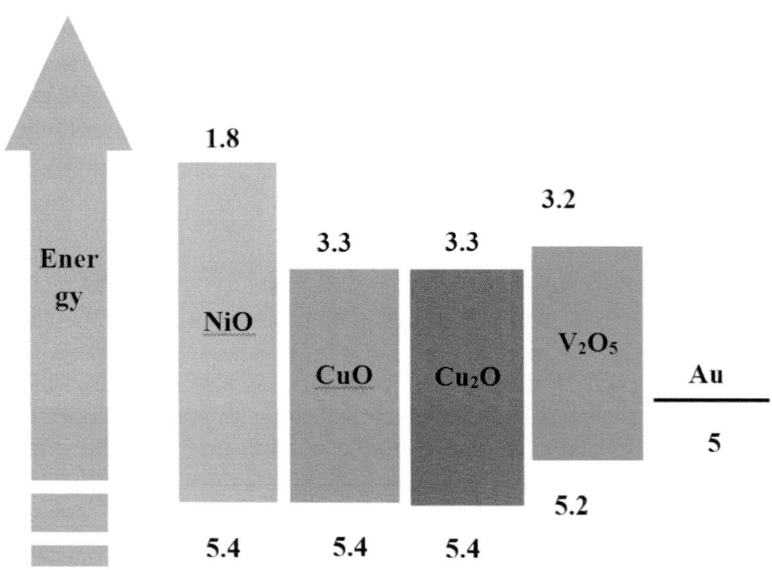

Figure 3. Energy diagrams of different MOs HTL layers used in PSCs.

Besides this, MOs are also utilized for passivation of PV devices. In order to reduce surface recombination losses in the device, the passivating layer is added. For high-efficiency solar cells, this passivation layer saturates the dangling bonds on the surface, lowering the surface state density and assisting in the reduction of recombination losses [28]. A variety of materials, including SiO_2 and Al_2O_3, are employed, along with different deposition methods like chemical vapour deposition, atomic layer deposition, and thermal oxidation. The most advantageous material for passivation is SiO_2, which has demonstrated superior passivation for both *n*- and *p*-type Si wafers [29].

3. Fuel Cells

MOs are critical components in the design of fuel cells, particularly solid oxide fuel cells (SOFCs). MOs of various compositions are employed in practically all aspects of SOFCs, including anode, electrolyte, and cathode materials, and are added to electrode materials in other fuel cells. Because of their ability to transfer charge with adsorbed molecules, MOs are used as electrocatalysts in fuel cell applications [30]. Transition MOs (TMO), in general, have high catalytic activity in the oxidation reduction process (ORR). Numerous TMO compounds currently exist and a considerable proportion of these take on a solid crystalline form under certain environmental conditions. The range of uses for fuel cells shows how versatile they are. They may be used to power various forms of transportation as well as vehicles such as cars, buses, drones, and even spacecraft. Research and development efforts integrating fuel cells into grid systems enable a smooth connection between storage and renewable energy sources. They also exhibit the potential for on-site, efficient, and clean electricity generation for residential and commercial applications. Different kinds of fuel cells have been produced, with differences in cost, applications, efficiency, and operating temperature. They are divided into six main groups according to the fuel and electrolyte that are utilized. The operation of the solid oxide fuel cell is depicted in Fig. 4.

Because of their superior performance, Pt-based catalysts are widely recognized as the most widely utilized electro catalysts in fuel cells [31]. Nevertheless, Pt-based catalysts are expensive, hard to come by, and poisonous during the oxygen reduction and methanol oxidation reactions (ORR). Therefore, in order to achieve comparable or better performance without the drawbacks of utilizing Pt-based catalysts in fuel cells, such as direct methanol fuel cells, MO nanosheets are being explored in conjunction with other materials. Because MO nanosheets enhance fuel cell performance, they are being investigated for potential application as electrocatalysts. It has been discovered that a Nickel oxide (NiO) nanosheet with numerous restricted oxygen vacancies can boost the electrocatalytic efficiency of MOR in direct methanol fuel cells. In contrast to bulk NiO and NiO nanosheets with low oxygen vacancies, abundant oxygen vacancies in the nanosheets demonstrated superior electrocatalytic performance

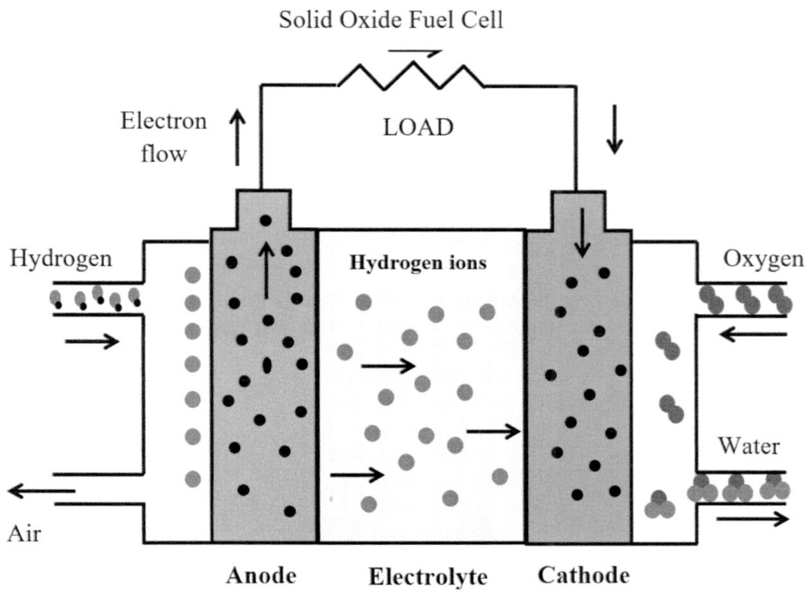

Figure 4. Schematic diagram of solid oxide fuel cell.

[32]. Therefore, it offers opportunities for use as dependable electrocatalysts for MOR. Improved ORR performance, high power density, and high durability were seen in Pt-TiO_{2-x} nanosheets with oxygen deficits as compared to anatase-TiO_2 nanosheets. There is a titanium cobalt oxide nanosheet-based platinum electrocatalyst for effective ORR in polymer electrolyte membrane fuel cells. The large specific surface area of the Co-doped TiO_2 nanosheets was attributed to their tubular form in the manufactured samples. Strong and dependable assistance was given by the suggested electrocatalyst to allow for the high dispersion of Pt nanoparticles [33]. Despite the fact that cerium doped co-oxide is a great catalyst with a large oxygen storage capacity and that it might be added to increase OER activity. The suggested $Co_{3-x}Ce_xO_4$ electrocatalyst's structure accelerated the synthesis of intermediates and enhanced electronic transport between Co and Ce, both of which raised OER activity [34]. Overall, the results of using MO nanosheets as a substitute for Pt-based electrocatalysts in different fuel cell systems are encouraging.

4. Batteries

Batteries are essential for energy storage and are used in many different applications, such as electric vehicles, renewable energy systems, and portable electronic devices. Chemical potential energy, which is transformed into electrical energy when needed, is stored in batteries. One or more electrochemical cells, each with an anode, cathode, and electrolyte, make up the fundamental parts of a battery. There are the following two types of batteries: primary and secondary. Primary batteries are for one-time application, as they have irreversible chemical reactions [35,36]. The price of these batteries is low and their application is simple for low-power devices. A few examples are as follows: zinc–carbon and zinc manganese alkaline batteries [37]. The specific applications are watches and torches. On the other hand, the application of secondary batteries is multiple times, as they are chargeable. Industries and automobiles are the main application of these batteries, where the value of the initial current should be high. However, the cost and the maintenance are the main disadvantages. The charging and discharging of the nickel battery is explained in detail, with the help of Fig. 5.

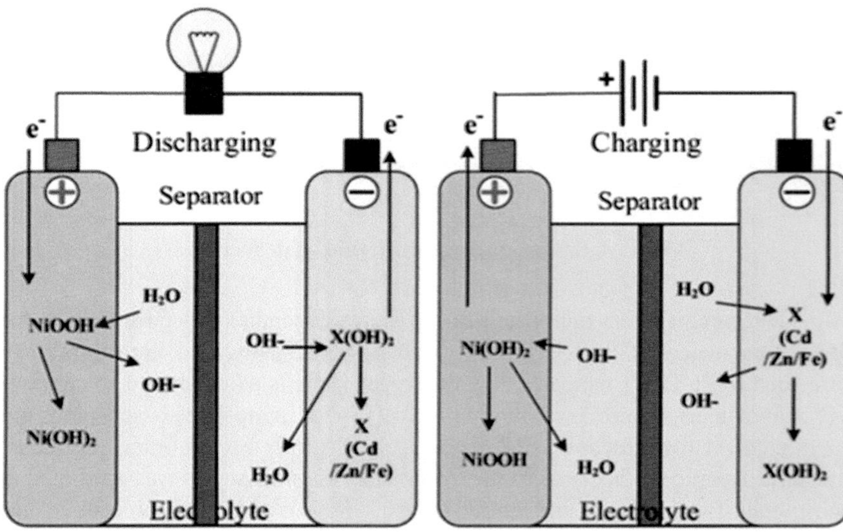

Figure 5. Schematic diagram of Ni battery.

The secondary battery (rechargeable) is growing day by day, due to an increase in the usage of portable devices, such as laptops, cell phones, notebook computers, etc. The traditional battery that is used in electric vehicles has a minimum life cycle of 1000 cycles, whereas a secondary (rechargeable) battery used in earth satellites have a life near 20,000 cycles. Secondary batteries come in several varieties, each with a unique set of features and useful uses. Because of its high energy density, extended longevity, and lightweight construction, the most common batteries used from practical aspects are Zn-ion, Li-ion, Na-ion, Li-S etc. The essential elements of the battery are electrode and MOs are frequently utilized as cathode materials in a large number of rechargeable batteries. During cycles of charging and discharging, MOs engage in reversible redox processes that enable them to store and release electrical energy. High energy density is a common characteristic of MOs, which is essential for extending battery life and enhancing battery performance in general. Literature work shows that researchers improvised the performance by implanting or diffusions metal ions into an battery. Liang et al. [38] developed a Zn-ion battery with a Ni-Mn nanosheet cathode with rich oxygen defects to improve conductivity, redox reaction activity, and provide sufficient ion diffusion channels for the electrochemical process. Similarly, Lu et al. and Kokits et al. [39,40] developed a novel electrode material consisting of 3D nanocubes of Zn hexacyanoferrate (ZnHCF) enveloped in MnO_2 nanosheet for rechargeable Zn-ion batteries, especially for wearable applications. Further, to improve the performance of Zn-ion batteries, Li et al. [41] developed Ni-doped Co-oxide nanosheets loaded onto a carbon-fiber paper substrate to be utilized as the air cathode. Ni-doped Co oxide nanosheets were utilized in the cathode for oxygen reduction reaction, achieved a 20% higher power density and can be used as a portable energy storage device.

Lithium-ion batteries frequently use MOs such as lithium iron phosphate ($LiFePO_4$), lithium cobalt oxide ($LiCoO_2$), and lithium nickel cobalt manganese oxide ($LiNiCoMnO_2$, NMC). Various MOs can be used, depending on the battery's particular needs. Zhang et al. [42] synthesized vanadium (III) oxide nanoparticles on conductive nitrogen-doped carbon-coated nanosheet array as a electrode to improve the performance of Li-ion batteries. The researchers enhanced the electrochemical performance due to uniform hierarchical structure, which increased the conductivity and reduced the volume expansion of the electrode. Another approach for improving the performance of Li-ion battery was performed by Li et al. [43], wherein they successfully synthesized 2D $LiNi_1/3Co_1/3Mn_1/3O_2$ nanosheets to serve as an active electrode in the battery. Moreover, this 2D or 1D configuration improved chemical reactivity due to the confined interlayers, higher surface area, and plentiful defect-rich edges which in turn improves the performance of the battery.

Sun et al. [44] synthesized 2D amorphous FeO_x nanosheets for utilization as an anode for Ni-ion batteries. Upon electrochemical testing, the amorphous FeO_x nanosheets exhibited speedier electron transfer reaction than crystalline FeO_x nanosheets, resulting in improving the specific capacitance, long-term cyclic stability, and good rate capability, making them highly appropriate for utilization in batteries. While other researchers proposed a simple and easy method for synthesizing a 2D amorphous MO nanosheets. Rubio et al. [45] synthesized Iron oxide-iron sulfide hybrid nanosheet anodes for the utilization in Na-ion batteries. Wang et al. [46] proposed the utilization of anatase TiO_2 nanosheets as the substrate in sulfur electrodes for Li-S batteries. The oxygen-deficient structure of anatase TiO_2 nanosheets boosted the electrochemical performance of the battery. The Na-S battery is a molten salt battery. The energy density, efficiency, charge/discharge, and cycle life are high, and the cost is low, as it can be produced from cheap materials [47]. On the other hand, maintaining the temperature between 300 and 350 °C, and the requirement of a permanent heat source that keeps the temperature within this limit and highly corrosive nature of the sodium polysulfide, and the increasing size of the cells, make it more uneconomical. Mehra et al. and few other researchers [48–50] invented the micro heat engines are capable to provide higher energy densities and maximum power output same as battery. The basic working principle of Micro heat engines is same as power plants. Basic process is to converts the chemical energy of the fuel into thermal energy through combustion, and then convert thermal to mechanical power by driving a heat engine. Although due to high cost of construction it's not used at commercial level. Recently, Betavoltaic systems are alternative energy technologies that promise extended battery life and power density over current technologies [51]. This system contains power sources of particles which convert high-energy beta particles emitted from the decay of radioactive isotopes into electrical current. Although the first photovoltaic battery achieved conversion efficiency is of around 4% and had an expected lifetime of almost 10 years. Recently, Olsen et al. reported 63Ni isotope-based betavoltaic cells which can provide kinetic energy up to 67 keV and have a half-life of 92 years [52].

Now a days, hybrid storage technologies having the complementary characteristics of both SC and rechargeable batteries has been established to store the energy [53–55]. Also, few more researchers reported a solar wind hybrid power unit which consists of a rechargeable battery and a fuel cell and are capable to use wind and solar resources for futuristic energy storage systems [56–59].

5. Supercapacitors

Supercapacitors are another energy storage devices that fill the void left by conventional capacitors and batteries. They are sometimes referred to as ultracapacitors or electrochemical capacitors. Through the electrostatic separation of charges, they store and release energy, offering great power density and quick energy bursts. The electrostatic separation of charges at the interface between an electrolyte and a high-surface-area electrode material is how supercapacitors store energy. Supercapacitors store energy electrostatically, which results in faster charge and discharge cycles than batteries, which depend on chemical reactions to store energy. Supercapacitors use electrical double-layer capacitance (EDLC) and pseudocapacitance as their two primary energy-storage methods. Ions are held in a double layer at the electrode-electrolyte interface, which is the site of EDLC. Faradaic redox processes at the electrode cause pseudocapacitance, which results in extra energy storage. High-surface-area materials, including conductive polymers or activated carbon, are used as electrodes in supercapacitors. At the electrode-electrolyte interface, a larger quantity of charge-carrying ions can be stored due to the increased surface area. Moreover, MOs are used as supercapacitors' electrode materials, increasing the devices' ability to store energy. Based on the energy storage technique, supercapacitors can be divided into two types: pseudocapacitors or EDLC (Fig. 6).

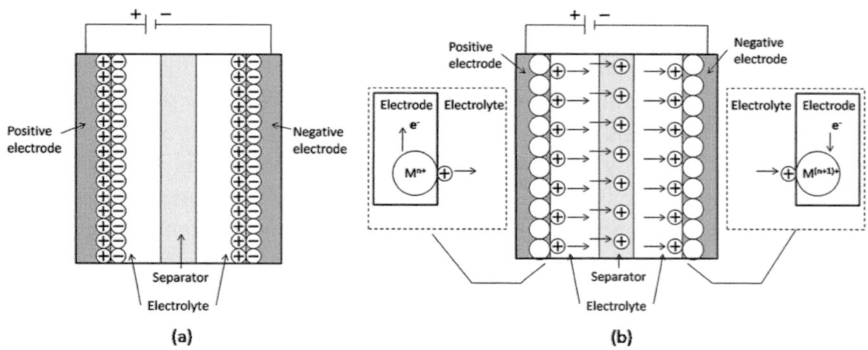

Figure 6. Schematic diagram of supercapacitors: (a) EDLC; (b) pseudocapacitor, M represents the metal atom.

The choice of MO can have an impact on a supercapacitor's energy density, power density, and cycling stability, among other performance characteristics. Conducting polymers, IrO_2, RuO_2, and NiO, etc. electrode materials have been demonstrated to increase device efficiency while lowering cost [60,61]. One common MO utilised in supercapacitors is ruthenium oxide, especially in the electrochemical capacitors called pseudocapacitors. Because of the reversible faradaic redox processes occurring at the electrode-electrolyte interface, it displays a high specific capacitance. Another MO that's frequently utilised in supercapacitors is manganese dioxide. It is cheap, plentiful, and safe for the environment. Electrical double-layer capacitance and pseudocapacitance work together to power MnO_2-based supercapacitors. The pseudocapacitive behaviour of supercapacitors is attributed to the use of nickel oxide as an electrode material. High energy density has been investigated as a potential application for NiO-based supercapacitors. The pseudocapacitive characteristics of cobalt oxide, more especially cobalt(III) oxide (Co_3O_4), are being studied in supercapacitors.

Co_3O_4 based electrodes have the potential to provide strong electrochemical performance, which raises the supercapacitor's total energy storage capacity. In addition to these, combinations of various MOs or mixed MOs are being investigated to produce synergistic effects and enhance the supercapacitor electrodes' overall performance. Supercapacitors' capacity to store electrical charge via both electrostatic and faradaic processes is facilitated by the use of MOs in them. The goal of this field's research is to improve supercapacitors' overall performance, cycling stability, and specific capacitance so they can compete more effectively with more established energy storage technologies across a range of applications. Through the creation of several redox processes, quaternary MOs improve the electrochemical characteristics of employed electroactive materials, hence increasing their energy storage capacity. The best electrocapacitive materials for battery-type electrodes were found to be quaternary

$NiCoMoFeO_x$, $NiCoMoZnO_x$, $NiCoMoAlO_x$, and $NiCoMoCuO_x$ materials, which were synthesized using a hydrothermal process and had attractive shapes [62].

MOs in supercapacitors have a bright future ahead of them as long as scientists keep discovering and developing new ways to store energy. A number of developments and trends are expected. The goal of ongoing research is to increase supercapacitors' energy density. This includes creating novel MO materials or enhancing the specific capacitance and overall energy storage capacity of already-existing ones. Such as graphene, nanotubes, and MO nanoparticles, are anticipated to proliferate. By increasing electrode surface area, nanostructuring can enhance charge storage and accelerate charge/discharge rates. In addition, there will probably be more research done on composite materials and mixed MOs. Combining various MOs or incorporating them with other materials can have synergistic effects that improve supercapacitors' overall electrochemical performance. MO materials with better structural and electrochemical properties can be produced with the help of synthesis technique advancements such template-assisted methods, hydrothermal methods, and sol-gel processes. The commercialization and broad usage of MO-based supercapacitors are predicted to increase with more research and the identification of promising materials, which will support the expansion of the energy storage market.

6. Issues, Challenges, and Future Directions

Although MOs have significantly advanced many fields, there are potential drawbacks as well. For MOs to reach their full potential and be addressed, it is imperative to comprehend these obstacles. Furthermore, some intriguing future applications for MOs are suggested by ongoing research. Because of their special qualities and wide range of uses, compounds made of MOs have drawn a lot of interest among the different types of nanomaterial's. For them to be even more reliable and effective in real-world applications, a number of issues and restrictions still need to be resolved.

(i) Challenges

1. **Synthesis Regulate and Reproducibility:** It can be difficult to precisely regulate the size, shape, and composition of MO nanoparticles, which can have an impact on the particles' performance and characteristics. For scalability and useful applications, reproducibility in the synthesis processes is essential. During catalytic processes, certain MOs may experience problems including agglomeration, sintering, or phase transitions that compromise their stability and endurance. Although having the capacity to fine-tune features is advantageous, controlling and optimizing these properties exactly for certain applications can also be difficult.
2. **Impact on the Environment and Health:** Some MOs, especially those that contain heavy metals, may be hazardous, which presents challenges

for the environment and human health. One of the main challenges is comprehending and reducing the possible environmental impact of MO nanoparticles.

3. **Device Integration:** Compatibility and interface issues must be resolved for MOs to be effectively incorporated into useful devices like solar cells, sensors, and batteries. It is crucial to provide long-term stability and functionality in practical settings. Certain MOs may have problems with affordability and resource availability, particularly those that incorporate pricey or uncommon elements.
4. **Elevated Temperature of Battery:** High-energy-density rechargeable batteries, especially Li-ion battery need to work under a moderate temperature ranging from −20 to 60 °C, ideally within a more restrictive temperature range between 20 and 40 °C. The elevated temperature accelerates the risks of overheating and thermal runaway if they work under high temperatures for a long duration, however, a low ambient temperature also threatens the working performance and safe operation of battery because the diffusion kinetics of Li^+ ions becomes slow and it degrades the performance in terms of both energy power and energy density. Thus, battery cooling at high temperatures and battery heating at low temperatures are simultaneously required to achieve highly efficient and safe operation of battery under various climate conditions.

(ii) Prospective Uses

1. **Advanced Energy Storage:** As the need for effective energy storage grows, MO nanoparticles are anticipated to be essential to the development of high-performance batteries and supercapacitors. The development of flexible and wearable electronic devices appears to be promising when it comes to MOs with distinct optical and electrical properties. In the world of medicine, MO nanoparticles are being investigated for targeted medication delivery, imaging, and therapies. In the future, MO-based supercapacitors could be widely used and commercialized, which would advance energy storage technology.
2. **Smart Coatings:** Self-cleaning surfaces, anti-corrosion coatings, and responsive materials are just a few of the uses for MO nanoparticles in smart coatings. The goal of ongoing research is to increase the stability and efficiency of MO-based parts used in photovoltaic systems, such as perovskite solar cells. Because of their special electrical characteristics, MOs are being investigated for their potential in quantum computing and nanoelectronics applications. Some MOs have the ability to fight bacteria, which makes them good choices for use in antimicrobial coatings and medical equipment. Because they break down pollutants in the presence of sunshine, MO nanoparticles—in particular, photocatalytic ones—have potential uses in the purification of water.
3. **Thermal Management System:** Thermal management systems have

been developed in the past decades to regulate battery temperatures within an optimal range. Liquid cooling method commercial used to dissipate the heat at the expense of bulky structure and extra battery bankup of electricity consumption. In recent years, another method has been used to optimized the battery temperature by utilizing the large endothermic heat enthalpies of these materials during the phase transformation from solid to liquid states like 220–251 kJ kg^{-1} for paraffin wax melting. This method is effectively used due to its low power consumption and excellent temperature uniformity. Further liquid–gas phase-change processes have been used since it show much higher heat dissipation capacities due to the high latent heats compare to solid liquid phase process. For the past few years, evaporating-based systems were proposed for specific hybrid electric vehicles (HEVs), where batteries were submerged in liquid fuels to prevent overheating and thermal runaway. The liquid fuels are heated to vaporization by absorbing the heat from batteries with increasing temperature; thereafter, the warm vapor is used to fuel the engine to achieve higher combustion efficiency. Recently, evaporation-based cooling system was reported to control the temperature of battery by using superabsorbent polymers to store liquid water and was fixed on the external surface of battery [63, 64]. In this method, water vapor is released from the polymers and then diffuses to the ambient air, removing a large amount of heat from the batteries to the surroundings.

7. Conclusion

Applications for MO thin films in energy harvesting and energy storage have garnered a lot of interest. These films are the result of multiple deposition techniques. When it comes to enhancing the effectiveness and efficiency of energy harvesting and storage devices, nanostructured MOs have a lot of potential to help create more efficient and sustainable energy sources. They are extremely helpful for enhancing the performance and efficiency of different gadgets because of their special qualities. MO nanoparticles have a lot of potential, but there are still issues that need to be resolved, like scalability in synthesis, long-term stability, and commercialization barriers. Overcoming these obstacles creates chances for additional study and creativity in the area. The global drive for ecologically friendly and sustainable energy solutions is in line with the usage of MO nanoparticles in energy harvesting and storage. Their incorporation into gadgets advances the creation of greener and more effective technologies. The full potential of these materials for useful applications may be unlocked by future advancements in synthesis methods, device design, and integration. The goal of ongoing research is to overcome obstacles and realize these materials' full potential for useful applications. Even if there are still issues with cost, scalability, and overall system optimization, continued research and technical developments are paving the way for the eventual integration of MOs in supercapacitors, which will provide answers for a range of energy storage requirements. New MO nanosheets with superior electrochemical qualities, like faster charge/discharge rates, large capacity, and greater stability, are being developed by researchers. This includes creating brand-new nanostructures, like hybrid nanosheets, and refining the properties of the nanosheets through the application of innovative synthesis techniques. High-performance and ecologically friendly supercapacitors and batteries for a variety of applications are being developed thanks to ongoing developments in material science, synthesis methods, and knowledge of electrochemical processes. Because of their adaptability and simplicity of integration into a variety of device topologies, MO nanosheets are a promising material for the development of flexible and wearable energy storage devices, an area of significant interest in this field. All things considered, MO nanosheets are probably going to become more and more significant in the development of sophisticated electrochemical energy storage devices in the years to come.

References

1. M. Farahmand Habibi, M. Arvand, S. Sohrabnezhad, Boosting bioelectricity generation in microbial fuel cells using metal@MOs/nitrogen-doped carbon quantum dots, Energy, 223 (2021) 120103.
2. A.P. Khedulkar, V.D. Dang, A. Thamilselvan, R. Doong, B. Pandit, Sustainable high-energy supercapacitors: MO-agricultural waste biochar composites paving the way for a greener future, Journal of Energy Storage, 77 (2024) 109723.
3. J. Zhang, Q. Yang, C. Gao, X. Wang, X. Liu, Copper doped lanthanum hydroxide nanorods as a low temperature processable hole transport material for perovskite solar cells, Journal of Power Sources, 590 (2024) 233797.
4. Y. Li, S. Wu, C. Wang, D. Du, Y. Gu, L. Bi, Sc-doping strategy for $LaNi_{0.5}Fe_{0.5}O_{3-\delta}$ cathode to boost the performance of proton-conducting solid oxide fuel cells, International Journal of Hydrogen Energy, 50 (2024) 124–133.
5. X. Li, Y. Guo, T. Gao, H. Liu, C. Chen, J. Li, D. Xiao, Rational design and construction of iron oxide and titanium carbide MXene hierarchical structure with promoted energy storage properties for flexible battery, Journal of Colloid and Interface Science, 631 (2023) 182–190.
6. A. Pattnaik, J.N. Sahu, A.K. Poonia, P. Ghosh, Current perspective of nano-engineered MO based photocatalysts in advanced oxidation processes for degradation of organic pollutants in wastewater, Chemical Engineering Research and Design, 190 (2023) 667–686.
7. S. Singh, R. Garg, A. Jana, C. Bathula, S. Naik, M. Mittal, Current developments in nanostructurally engineered MO for removal of contaminants in water, Ceramics International, 49(5) (2023) 7308–7321.
8. X. Jin, H. Sun, Y. Tong, J. Zou, J. Zhao, C. Qin, Y. Wang, Z. Yan, Porous hollow CaO microsphere synthesized by the template-assisted approach for enhanced CO_2 capture, Fuel, 361 (2024) 130638.

9. F. Li, X. Zhang, S. Liang, M. Sun, X. Zhao, H. Chen, Y. Cuia, Facile and scalable synthesis of 2D porous Ni/C via a salt-template assisted approach for enhanced urea oxidation reaction and energy-saving hydrogen production, New Journal of Chemistry, 47(15) (2023) 7399–7409.
10. X. Luo, M. Yang, J. Bu, T. Chen, A.Y. Yi, W. Xu, Precise fabrication of nanostructured mixed MOs by the integration of nanoimprinting and sol–gel synthesis, Macromolecular Research, 31(8) (2023) 795–803.
11. T. Yıldız, N. Kati, B. Gül, Characterization of CuO doped CdO nanomaterials synthesized by sol gel spin coating and hydrothermal method, Materials Science and Engineering: B, 290 (2023) 116306.
12. S. Manisha Pathania, P. Chinnamuthu, D. Kumar, T. Kumar, V. Singh, R. Jha, J. Jire, L. Hmar, Wearable flexible memristor based on titanium dioxide (TiO_2)-Zinc oxide (ZnO) embedded in polyvinyl alcohol (PVA) matrix, Materials Science in Semiconductor Processing, 170 (2024) 107953.
13. B.A. Al-Mur, Green Zinc Oxide (ZnO) Nanoparticle synthesis using mangrove leaf extract from Avicenna marina: Properties and application for the removal of toxic metal ions (Cd^{2+} and Pb^{2+}), Water, 15(3) (2023) 455.
14. P.S. Chauhan, S. Kumar, A. Mondal, P. Sharma, M.N. Parekh, V. Panwar, A.M. Rao, A. Misra, Stacked vanadium pentoxide–zinc oxide interface for optically-chargeable supercapacitors, Journal of Materials Chemistry A, 11(1) (2023) 95–107.
15. M. Kim, J. Choi, W. Lee, Y.Y. Ahn, H. Lee, K. Cho, J. Lee, Performance of Magnéli phase Ti_4O_7 and Ti^{3+} self-doped TiO_2 as oxygen vacancy-rich titanium oxide anodes: Comparison in terms of treatment efficiency, anodic degradative pathways, and long-term stability, Applied Catalysis B: Environmental, 337 (2023) 122993.
16. L. Gao, Z. Wang, C. Qin, Z. Chen, M. Gao, N. He, Xi. Qian, Z. Zhou, G. Li, Preparation and application of iron oxide/persimmon tannin/graphene oxide nanocomposites for efficient adsorption of erbium from aqueous solution, Journal of Rare Earths, 38(12) (2020) 1344–1353.
17. A. Ihsan, S.T. Almutairi, H.M.A. Mahmoud, M.F. Warsi, S. Munir, A strategy to boost the electrochemical properties of Ag–Fe_2O_3 with intercalation of MXene hydrogel, Ceramics International, 49(23) (2023) 37650–37660.
18. F.H. Isikgor, S.T. Zhumagali, L.V. Merino, M.D. Bastiani, I. McCullouch, S.D. Wolf, Molecular engineering of contact interfaces for high-performance perovskite solar cells, Nature Reviews Materials, 8(2) (2022) 89–108.
19. J. Zhang, J. Mcgettrick, K. Ji, J. Bi, T. Webb, X. Liu, D. Liu, A. Ren, Y. Xiang, B. Li, V. Stolojan, T. Watson, S.D. Stranks, W. Zhang, Fast and balanced charge transport enabled by solution-processed MO layers for efficient and stable inverted perovskite solar cells, Energy & Environmental Materials (2023).
20. H.-S. Kim, C.-R. Lee, J.-H. Im, K.B. Lee, T. Moehl, A. Marchioro, S.J. Moon, R.H. Baker, J.H. Ho Yum, J.E. Moser, M. Grätzel, N.G. Park, Lead iodide perovskite sensitized all-solid-state submicron thin film mesoscopic solar cell with efficiency exceeding 9%, Scientific Reports, 2(1) (2012.
21. S. Muhammad, J. Arshad, S. Abu Bakar, H. Ahmad, S. Qaisar, Synergic effect of Zinc (Zn) precursors on the microstructural, optical and electrical properties of Zinc oxide thin films for solar cell applications, Optik, 206 (2020) 163750.
22. J. Qiu, Y. Qiu, K. Yan, M. Zhong, C. Mu, H. Yan, S. Yang, All-solid-state hybrid solar cells based on a new organometal halide perovskite sensitizer and one-dimensional TiO_2 nanowire arrays, Nanoscale, 5(8) (2013) 3245.

23. N.H. Shamsudin, S. Shafie, M.Z.A. Ab Kadir, F. Ahmad, Y. Sulaiman, S.A.M. Chachuli, M.C. Razali, Flexible back-illuminated dye sensitised solar cells (DSSCs) with titanium dioxide/silver nanoparticles composite photoanode for improvement of power conversion efficiency, Optik, 272 (2023) 170237.
24. P. Li, Y. Huang, Y. Zhang, L. Chen, Performance and mechanism of Al, N co-doped carbon/nano-TiO_2 photocatalytic oxidation for the removal of ammonia nitrogen and Ni/Co complexes from ternary precursor wastewater, Chemical Engineering Journal, 479 (2024) 147829.
25. A. Kumar, S. Singh, A. Yadav, Recent progress in inverted perovskite solar cells employing nickel oxide (NiO_x) as a hole transport materials, Materials Today: Proceedings, 46 (2021) 5827–5832.
26. M.A. Zahid, M.Q. Khokhar, Z. Cui, H. Park, J. Yi, Improved optical and electrical properties for heterojunction solar cell using Al_2O_3/ITO double-layer anti-reflective coating, Results in Physics, 28 (2021) 104640.
27. E.Y. Choi, J. Kim, S. Lim, E. Han, A.W.Y. Ho-Baillie, N. Park, Enhancing stability for organic-inorganic perovskite solar cells by atomic layer deposited Al_2O_3 encapsulation, Solar Energy Materials and Solar Cells, 188 (2018) 37–45.
28. M.Z. Rahman, S.I. Khan, Advances in surface passivation of c-Si solar cells, Materials for Renewable and Sustainable Energy, 1(1) (2012).
29. S.W. Glunz, F. Feldmann, SiO_2 surface passivation layers – A key technology for silicon solar cells, Solar Energy Materials and Solar Cells, 185 (2018) 260–269.
30. Y.N. Yusoff, N. Shaari, S.H. Osman, MO-based materials as an emerging platform for fuel cell system: A review, Materials Science and Technology, 39(16) (2023) 2363–2397.
31. X. Wang, Z. Li, Y. Qu, T. Yuan, W. Wang, Y. Wu, Y. Li, Review of metal catalysts for oxygen reduction reaction: From nanoscale engineering to atomic design, Chem, 5(6) (2019) 1486–1511.
32. W. Yang, X. Yang, J. Jia, C. Hou, H. Gao, Y. Mao, C. Wang, J. Lin, X. Luo, Oxygen vacancies confined in ultrathin nickel oxide nanosheets for enhanced electrocatalytic methanol oxidation, Applied Catalysis B: Environmental, 244 (2019) 1096–1102.
33. F. Yu, Y. Xie, L. Wang, N. Yang, X. Meng, X. Wang, X.L. Tian, X. Yang, Platinum supported on multifunctional titanium cobalt oxide nanosheets assembles for efficient oxygen reduction reaction, Electrochimica Acta, 265 (2018) 364–371.
34. L. Li, Z. Hu, L. Tao, J. Xu, J.C. Yu, Efficient electronic transport in partially disordered Co_3O_4 nanosheets for electrocatalytic oxygen evolution reaction, ACS Applied Energy Materials, 3(3) (2020) 3071–3081.
35. R.A. Powers, Advances and trends in primary and small secondary batteries, IEEE Aerospace and Electronic Systems Magazine, 9(4) (1994) 32–36.
36. R. Faria, P. Marques, R. Garcia, P. Moura, F. Freire, J. Delgado, A.T. de Almeida, Primary and secondary use of electric mobility batteries from a life cycle perspective, Journal of Power Sources, 262 (2014) 169–177.
37. X. Hu, A. Robles, T. Vikström, P. Väänänen, M. Zackrisson, G. Ye, A novel process on the recovery of zinc and manganese from spent alkaline and zinc-carbon batteries, Journal of Hazardous Materials, 411 (2021) 124928.
38. X. Liang, X. Zou, X. Ren, B. Xiang, W. Li, F. Chen, J. Hao, Q. Hu, L. Feng, Ultrathin nickel manganese nanosheets with rich oxygen-vacancy as a durability electrode for aqueous Ni/Zn batteries, Journal of Colloid and Interface Science, 578 (2020) 677–684.

39. L. Luo, Y. Liu, Z. Shen, Z. Wen, S. Chen, G. Hong, High-voltage and stable manganese hexacyanoferrate/zinc batteries using gel electrolytes, ACS Applied Materials & Interfaces, 15(24) (2023) 29032–29041.
40. O. Kokits, Y. Aniskevich, A. Mazanik, O. Yakimenko, G. Ragoisha, S.T. Myung, E. Streltsov, Origin of electrochemical activation on vanadium hexacyanoferrate cathode for aqueous zinc-ion battery, Energy Storage Materials, 63 (2023) 103017.
41. Y. Li, L. Cui, P. Da, K.W. Qiu, W.J. Qin, W.B. Hu, X.W. Du, K. Davey, T. Ling, S.Z. Qiao, Multiscale structural engineering of Ni-doped CoO nanosheets for zinc–air batteries with high power density, Advanced Materials, 30(46) (2018).
42. X. Zhang, L. Xun, S. Gao, Y. Xu, X. Cheng, H. Zhao, L. Huo, Facile synthesis of V_2O_3@N-doped carbon nanosheet arrays on nickel foam as free-standing electrode for high performance lithium ion batteries, Catalysis Today, 374 (2021) 117–123.
43. J. Li, X. Wang, J. Zhao, J. Chen, T. Jia, C. Cao, Porous lithium nickel cobalt manganese oxide hierarchical nanosheets as high rate capability cathodes for lithium ion batteries, Journal of Power Sources, 307 (2016) 731–737.
44. R. Sun, J. Gao, G. Wu, L. Peigen, G. Wenxin, Z. Huang, G. Jing-Jie, H. Yanmin, X. Zhenggang, L. Hai, C. Pei-Xin, Z. Xusheng, W. Yuen, Z. Genqiang, H. Xun, Amorphous MO nanosheets featuring reversible structure transformations as sodium-ion battery anodes, Cell Reports Physical Science, 1(7) (2020) 100118.
45. S. Rubio, R.R. Maça, G.F. Ortiz, C.P. Vicente, P. Lavela, V. Etacheri, J.L. Tirado, Iron oxide–iron sulfide hybrid nanosheets as high-performance conversion-type anodes for sodium-ion batteries, ACS Applied Energy Materials, 3(11) (2020) 10765–10775.
46. H. Wang, C. Fan, Y. Zheng, W.H. Li, S.Y. Liu, H.Z. Sun, J.P. Zhang, L.N. Sun, X.L. Xu, Oxygen-deficient titanium dioxide nanosheets as more effective polysulfide reservoirs for lithium-sulfur batteries, Chemistry – A European Journal, 23(40) (2017) 9666–9673.
47. H. Liu, X. Zhang, S. He, Y. Shang, H. Yuet, Molten salts for rechargeable batteries, Materials Today, 60 (2022) 128–157.
48. C.M. Spadaccini, A. Mehra, J. Lee, X. Zhang, S. Lukachko, I.A. Waitz, High power density silicon combustion systems for micro gas turbine engines, Journal of Engineering for Gas Turbines and Power, 125(3) (2003) 709–719.
49. S. Whalen, M. Thompson, D. Bahr, C. Richards, R. Richardset, Design, fabrication and testing of the P3 micro heat engine, Sensors and Actuators A: Physical, 104(3) (2003) 290–298.
50. A. Berni, Review of 'Wireless Ad Hoc and Sensor Networks' (1st ed.). By Houda Labiod. John Wiley & Sons, Inc. (2008), $165.00, ISBN: 1848210035, Queue, 7(1) (2009) 40–40.
51. M.V.S. Chandrashekhar, C.I. Thomas, H. Li, M.G. Spencer, A. Lal, Demonstration of a 4H SiC betavoltaic cell, Applied Physics Letters, 88(3) (2006).
52. L.C. Olsen, P. Cabauy, B.J. Elkind, Betavoltaic power sources, Physics Today, 65(12) (2012) 35–38.
53. M.T. Penella, M. Gasulla, A simple and efficient MPPT method for low-power PV cells, International Journal of Photoenergy (2014) 1–7.
54. M.T. Penella, M. Gasulla, Runtime extension of low-power wireless sensor nodes using hybrid-storage units, IEEE Transactions on Instrumentation and Measurement, 59(4) (2010) 857–865.
55. M. Alzaid, R. Alanazi, I.H. Alsohaimi, W.S. Mohamed, N.M.A. Hadia, M. Ezzeldien, M.Z. Iqbal, Hydrothermally assisted bimetallic transition metal sulfide as battery

grade electrode and activated carbon as capacitive electrode for hybrid energy storage devices, Diamond and Related Materials, 134 (2023) 109737.
56. S. Sambhi, H. Sharma, V. Bhadoria, P. Kumar, R. Chaurasia, G. Fotis, Technical and economic analysis of solar PV/diesel generator smart hybrid power plant using different battery storage technologies for SRM IST, Delhi-NCR Campus, Sustainability, 15(4) (2023) 3666.
57. P.K. Sharma, D.A. Kumar, P. William, D. Obulesu, P.M. Pandian, T.K.H. Khan, G. Manikandan, Energy storage system based on hybrid wind and photovoltaic technologies, Measurement: Sensors, 30 (2023) 100915.
58. Q. Gao, A. Bechlenberg, B. Jayawardhana, N. Ertugrul, A.I. Vakis, B. Ding, Techno-economic assessment of offshore wind and hybrid wind–wave farms with energy storage systems, Renewable and Sustainable Energy Reviews, 192 (2024) 114263.
59. S. Farah, G.B. Andresen, Investment-based optimisation of energy storage design parameters in a grid-connected hybrid renewable energy system, Applied Energy, 355 (2024) 122384.
60. Y. Liu, S.P. Jiang, Z. Shao, Intercalation pseudocapacitance in electrochemical energy storage: Recent advances in fundamental understanding and materials development, Materials Today Advances, 7 (2020) 100072.
61. H. Lichchhavi Lee, Y. Ohshita, A.K. Singh, P.M. Shirage, Transformation of battery to high performance pseudocapacitor by the hybridization of W18O49 with RuO_2 nanostructures, Langmuir, 37(3) (2021) 1141–1151.
62. L. Halder, A. Maitra, A.K. Das, R. Bera, S.K. Karan, S. Paria, A. Bera, S. Kumar, S.B.B. Khatua, Fabrication of an advanced asymmetric supercapacitor based on three-dimensional copper–nickel–cerium–cobalt quaternary oxide and GNP for energy storage application, ACS Applied Electronic Materials, 1(2) (2019) 189–197.
63. X. Fan, R. Zhang, S. Sui, X. Liu, J. Liu, C. Shi, N. Zhao, W. Hu, Starch-based superabsorbent hydrogel with high electrolyte retention capability and synergistic interface engineering for long-lifespan flexible zinc–air batteries, Angewandte Chemie, 135(22) (2023).
64. A. El Idrissi, B. Channab, Y. Essamlali, M. Zahouily, Superabsorbent hydrogels based on natural polysaccharides: Classification, synthesis, physicochemical properties, and agronomic efficacy under abiotic stress conditions: A review, International Journal of Biological Macromolecules, 258 (2024) 128909.

CHAPTER

6

A Gas Sensor Array Based on Metal Selenides

Sanju Rani[1*], Manoj Kumar[5*], Kuldeep Singh Gour[4], Gopal Rawat[1] and Vidya Nand Singh[2,3]

[1] School of Computing and Electrical Engineering (SCEE), Indian Institute of Technology Mandi, Kamand, Himachal Pradesh 175075, India

[2] Academy of Scientific and Innovative Research (AcSIR), Ghaziabad, Uttar Pradesh 201002, India

[3] Indian Reference Materials (BND) Division, CSIR-National Physical Laboratory, Dr. K.S. Krishnan Marg, New Delhi 110012, India

[4] Corrosion & Coating Group, Advanced Materials & Corrosion Division, CSIR-National Metallurgical Laboratory, Burma Mines, Jamshedpur 831007, India

[5] School of Basic and Applied Sciences, Institute of Integrated Learning and Management University, Greater Noida, Uttar Pradesh 201306, India

1. Introduction

Air pollution has become a more serious issue due to the recent and rapid rise of industrialization and urbanization. Among the most well-known pollutants are nitrogen dioxide (NO_2), acetone, hydrogen disulfide (H_2S), ammonia (NH_3) etc. The main cause of lung conditions like chronic bronchitis, emphysema, and respiratory irritation disorder is due to these gases, which have extremely negative consequences on human health [1,2]. As a result, to monitor these gases quickly and effectively, portable, affordable, and real-time technologies must be developed [3]. A gas sensor is a device that measures the concentration of different gases in the air. It is widely used in environmental monitoring, industrial safety, medical care, and conservation of the environment [4–6]. Because of their easy construction, low cost, straightforward operation, and compact size, resistance-type gas sensors are the most appealing and useful

*Corresponding authors: visanjun@gmail.com; manoj15111993kumar@gmail.com

among the different types of gas sensors for use in sensing poisonous analysts and explosive gases [7]. Metal oxide-based chemiresistive gas sensors have been extensively utilized to detect NO_2, H_2S, CO, and NH_3 with a high degree of sensitivity [3]. The slow response/recovery periods of these technologies create a significant barrier to their commercialization [8]. Due to their high operating temperatures and power consumption, the majority of MOS gas sensors are not well suited for integration [3].

Numerous promising materials, including polymers [9], metal-organic frameworks (MOFs) [10], 2D transition-metal dichalcogenides (TMDCs) [11], and graphene [12], have been studied recently in an effort to produce gas sensors that operate at lower temperatures. In particular, transition metal dichalcogenides (TMDCs) semiconducting materials have drawn a lot of attention due to their exceptional transport capabilities and layered structure [6]. Since metal chalcogenides have important electrical and catalytic properties, they have become an attractive choice for use in gas and humidity sensors. Metal chalcogenides' size and engineering surface properties have a significant impact on their chemical, physical, and sensing properties [13].

For use in gas sensor applications, a range of metal chalcogenides, including CdS [14], MoS_2 [15], WS_2 [16], SnS_2 [17], GeSe [18], PbS [19], InSe [13], WSe_2 [2], Bi_2Se_3 [20], $MoSe_2$ [21], $SnSe_2$ [22], CdSe [23], SnSe [24], and ZnSe [25], are reported. Among these materials, we discussed in this chapter only the gas-sensing properties of selenide-based materials such as InSe, WSe_2, $MoSe_2$, SnSe, and $SnSe_2$, respectively. While transition metal di-selenides have been investigated for many years, new methods for realizing ultra-sensitive and quick-response sensing devices have been made possible by advancements in nanoscale material synthesis and the manufacturing of effective nanoelectronics circuits [26].

Recently, indium selenide (InSe), a common III–VI layer chemical, has been discovered. InSe has an n-type conducting nature, and its bandgap ranges from 2.11 eV for monolayer to 1.26 eV for bulk [27], has gained significant interest because of its unique chemical and physical characteristics [28]. Furthermore, based on the charge-transfer mechanism, density functional theory (DFT) studies have already shown that 2D InSe has a superior gas-sensing capability for identifying open-shell molecules (NO_2 and NO) [29]. However, due to their extraordinary sensitivity to the surrounding environment, two-dimensional (2D) TMDCs semiconducting materials of the type MX_2, where M = Mo, W, and X = S, Se, are enticing materials to be studied in gas-sensing applications. These materials' high surface-to-volume ratio contributes to their remarkable response as well [2]. $MoSe_2$ and WSe_2 are both porous materials that can be exfoliated into two-dimensional layers with a single-unit cell thickness because of their strong in-plane bonding and weak out-of-plane interactions [26]. P-type conductivity is found in both materials. The bandgaps of P-type WSe_2 are 1.2 eV in bulk and 1.7 eV in monolayer [30], while $MoSe_2$ has a bandgap of 1.5 eV [6]. Another material, SnSe, is a significant II–VI group 2D semiconductor material that can absorb the majority of the light energy in the solar spectrum. Its direct band gap is 1.30 eV, while its indirect band gap is 0.90 eV [31]. Sn and Se atoms are organised in a zigzag pattern in each molecular layer, giving it the same layered structure as black phosphorus [32]. Additionally, compared to other nonmagnetic gases (such as CO_2, H_2, H_2S, and CH_4), NO_2 has a greater affinity due to the spin polarisation of the SnSe super cell and its paramagnetic nature [33]. This makes it appropriate for SnSe to detect NO_2 gas [34]. $SnSe_2$ is an anisotropic binary layered material stacked in the Se–Sn–Se configuration. It has a hexagonal structure of type CdI_2 [35]. Recent research has revealed the material's rich properties and potential applications. Its tunable narrow band gap is about 1-2 eV [36]. Additionally, Sn and Se are both environmentally pleasant as well as common elements; they are members of the VIA and IVA families, respectively [37]. $SnSe_2$ is an attractive candidate for practical sensing applications due to its straight band-gap nature, huge surface area, and inexpensive cost [38]. When compared to other 2D nanomaterials, $SnSe_2$ demonstrated superior gas NH_3 sensing performance [38]. The sensing behaviour of all these selenide materials is good. With the use of these materials, gas sensing applications can improve their sensitivity towards dangerous gas. The all selenides material-based gas sensors shown in Figure 1.

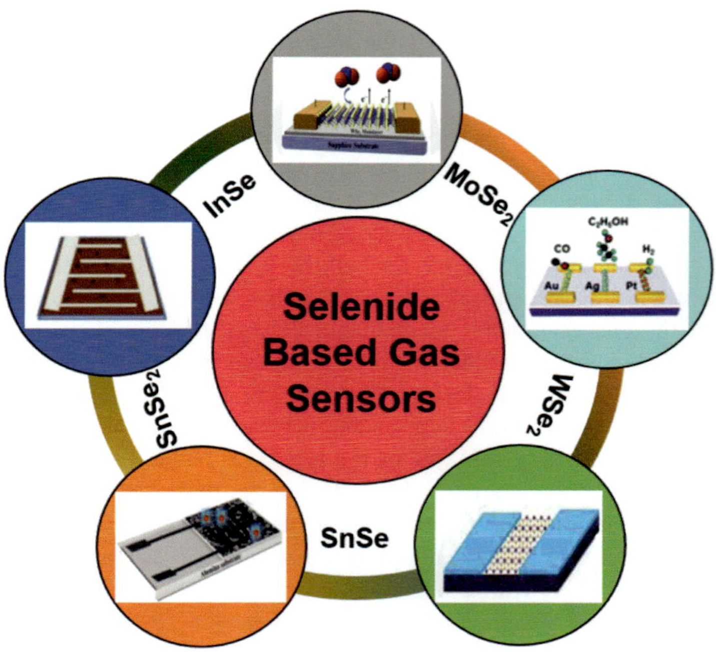

Figure 1. Shows selenide-based gas sensors.

2. Synthesis and Device Fabrication

i. Material Synthesis

Both chemical and physical methods are used to synthesise the substance. Many chemical and physical techniques have been used by researchers worldwide to produce materials based on selenides. One-pot, solvothermal, and other chemical techniques, as well as other physical techniques involving solid state reactions, are employed in the synthesis of materials based on selenides. These procedures that produce selenide-based compounds are briefly described in the sections that follow.

(a) Synthesis of InSe nanoparticles

An easy one-pot method was used for making InSe nanoparticles (shown in Figure 2). The mole ratio was chosen to be 1:1 in a typical experiment. 10 mL of OLA was added as a solvent, surfactant, and reductant to decrease selenium after approximately 0.3948 g (5 mmol) of indium chloride and 1.1059 g (5 mmol) of selenium powder were added to a 50 mL three neck flask. After that, using nitrogen gas and vigorous magnetic stirring, the liquid was heated rapidly to 200 °C. The mixture, which seemed homogeneous at first, changed colour during the heating process from black to yellow to dark brown after the temperature reached 200 °C. Aliquots were taken at various times at this temperature and left to naturally cool to 80 °C. To get the particles to flocculate, ethanol was used to wash the samples. After that, the particles were gathered by centrifuging for 10 minutes at 3000 rpm. To get dark brown powders, the products were dried at room temperature for 12 hours after being repeatedly washed to get rid of any remaining impurities. 10, 30 and 60 minutes were the time periods that were utilised in the synthesis of InSe nanoparticles [39].

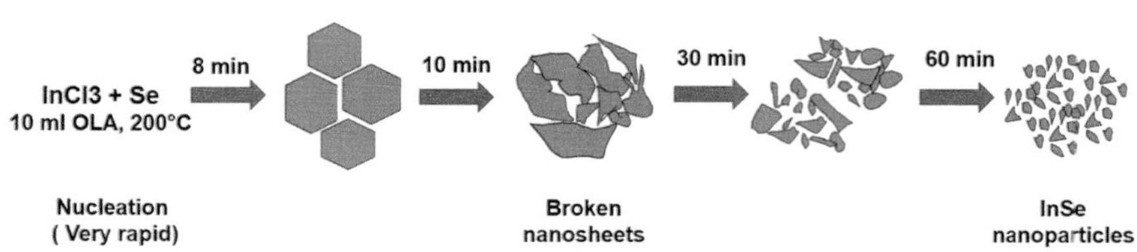

Figure 2. Proposed growth mechanism for the InSe nanoparticles with the images [39].

(b) Synthesis of WSe_2 nanorods

The solvothermal approach was used to synthesise WSe_2 (shown in Figure 1). Researcher was prepared two solutions. In first solution, 70 ml of ethylene glycol $(CH_2OH)_2$ was mixed with 2 g of selenium (Se) powder. The mixture was then placed in a Teflon-lined autoclave at 180 °C for 5 hours. In second solution, the mixture of 10 ml of distilled water and 40 ml of ethylene glycol was prepared and after that 2 g of sodium tungstate (Na_2WO_4) mixed in this mixture. To each solution, they added 2 ml of H_2O_2. In a 60 ml solution of distilled water, they blended 2 g of PVP and 3 g of urea properly. This mixture was put in two parts, half to the Se solution and half to the other. The other was then gradually filled with drops of sodium tungstate solution. Next, sonicate the mixture for 1 hours. The solution should then be vigorously stirred for two hours. In a 200 ml Teflon-lined autoclave, this mixture was inserted and heated to 200 °C for a period of 14 hours. After that, a brown solution with black precipitates was produced. After centrifuging the resultant material, it was thoroughly cleaned using ethanol and distilled water. Following washing, place the black precipitates in the petri dish and let the product dry for 6 hours at 60 °C in the oven [40].

(c) Synthesis of $MoSe_2$ nanoflower

During the synthesis, 10 ml of hydrazine hydrate ($N_2H_4 \cdot H_2O$) and 60 ml of deionized water were added together with 0.483 g of $Na_2MoO_4 \cdot {}_2H_2O$ and 0.316 g of Se powder. The mixture was stirred constantly to form a red solution, which was then adjusted to a pH of about 12 by adding a NaOH solution. The liquid was vigorously stirred, then transferred to a 100 ml Teflon-lined stainless autoclave. It was then tightly sealed, kept at 200 °C for 48 hours, and allowed to naturally cool to ambient temperature. After that, the suspension underwent several filters and washes with pure ethanol and distilled water. Ultimately, black $MoSe_2$ was produced by drying the prepared powder in the air for 12 hours at 70 °C [41].

(d) Synthesis of $SnSe_2$ nanosheets

A typical $SnSe_2$ nanoflower making consisted of dissolving 0.5 mmol of $SnCl_2$ in 20 ml of oleic acid and aggressively stirring the mixture for 30 minutes. After adding 1 mmol of Se metal powder, the mixture was stirred for 30 minutes. After that, the mixture was shifted to a Teflon tube that had a stainless-steel autoclave lining it, and it was kept in furnace for 48 hours at 220 °C. Upon reaching room temperature, the autoclave naturally cooled down after 48 hours. Following multiple ethanol washes, a black precipitate was collected, then dried it for 2 hours at 60 °C. Excess ethanol and oleic acid were combined to create a mixed solvent that was utilised to create $SnSe_2$ nanosheets [42].

(e) Synthesis of SnSe nanorods

In a stoichiometric ratio of 1:1, Sn (purity $\geq$ 99.99%, α-Aesar) and Se (purity $\geq$ 99.999%, CERAC) were combined and the mixture was sealed in a quartz tube at 10^{-5} torr pressure. The sealed tube was then placed in a furnace that ramped up to 1173 K at a rate of 2 °C per minute, and it was held there for 10 hours before natural cooling. The resulting powder was then crushed and ground into smaller nanorods [43].

ii. Thin Film Deposition

Chemical and physical deposition techniques are two types of thin film deposition techniques. Researchers from all over the world have deposited selenide based thin films using a variety of chemical and physical deposition techniques. Researchers have used these methods to deposit selenide based thin films for gas sensing applications. The hydrothermal method, spin coating, etc., are chemical methods for depositing selenide based thin films. Thermal evaporation, pulsed-laser deposition (PLD), co-evaporation technique, chemical vapour deposition (CVD), and other physical techniques are physical vapour deposition techniques. The following sections provide a brief description of these techniques to produce selenide based thin films.

(a) Deposition of InSe thin film

InSe films were produced via pulsed-laser deposition (PLD) in a vacuum on semi-insulating InP (111) and insulating Al_2O_3 (001) substrates. The base vacuum pressure inside the chamber was maintained around ~10^{-6} Torr. A KrF excimer laser (~248 nm) was used to ablate an InSe target at a repetition rate of 10 Hz. During deposition, the substrate temperature is held constant at 350 °C. After 30 minutes of in-situ annealing at 450 °C, the films naturally cooled to ambient temperature [44].

(b) Deposition of $MoSe_2$ thin film

The CVD technique has been developed to synthesise monolayer $MoSe_2$. The alumina boat was filled with molybdenum oxide (MoO_3) (99%, Sigma Aldrich) and selenium pellets (99.9%, Sigma Aldrich) as precursors for selenium and molybdenum, respectively. Face down in the boat was a pristine Si wafer with a top layer of SiO_2 measuring 275 nm. The boat containing the precursor and substrate was situated in the hot zone, or the middle of the reactor, inside a fused quartz tube. $MoSe_2$ triangle domains or films were produced by increasing the furnace temperature to 750 °C with a heating ramp of 50 °C/min and holding it there for 20 minutes. The furnace was allowed to cool naturally after growth. A 50 standard cubic centimeters per minute (sccm) mixture of argon and hydrogen (15% hydrogen) was utilised as the reducing environment during the deposition process and as the carrier gas during the entire process. Growth continued in the presence of atmospheric pressure [45].

(c) Deposition of WSe_2 thin film

The following is a description of the standard synthesis process. 60 ml DMF was used to dissolve 0.20 g of sodium borohydride ($NaBH_4$), 9.2 mmol of selenium powder, and 4 mmol of sodium tungstate dehydrate ($Na_2WO_4{\cdot}_2H_2O$). For approximately one to two hours, the mixture was constantly agitated until a dark red solution was achieved.

After that, a 100 ml Teflon liner's inner wall was pressed against an angled position containing a thoroughly cleaned quartz substrate. The Teflon liner was filled with the prepared precursor solution and 10 ml of deionized water, and the stainless-steel autoclave was then loaded with Teflon liner and heated to 200 °C for 48 hours. The quartz substrate was removed from the autoclave when it had naturally cooled, washed with deionized water, and allowed to dry in room temperature [30].

(d) Deposition of $SnSe_2$ thin film

Using a co-evaporation technique from single element sources (Sn, Se), thin films of $SnSe_2$ was produced using a molecular beam epitaxy deposition device. The evaporation system's background pressure was 2.5×10^{-5} Pa. Sn, Se, and the substrate (Glass +PI) were heated to 1,100 °C, 215 °C, and 250 °C, respectively. There was a 20-minute deposition [38].

(e) Deposition of SnSe thin film

The typical solid-state reaction of the elements was used to create SnSe powder. After being sufficiently crushed, the resulting powder was used as a deposition source material. Using the thermal evaporation method, SnSe nanostructured film was formed on soda-lime glass for 2 min at a low tension voltage of 20 V and a low tension current (LTC) of 90 A. Figure 1 depicts the entire procedure [46].

iii. Choice of Electrode

The materials and sensor device design have a big impact on the sensor performance. Since most semiconducting selenide materials work at room temperature, these types of gas sensors are not dependent on heaters to maintain their working temperature. The electrode is deposited on one side of the substrate, whether it top or back of the film. Gold, platinum, silver, and aluminium are used to deposit electrodes, and any insulating material, such as glass, silicon dioxide, or alumina, is typically used as the substrate. This kind of gas sensor uses less power because is not requirement of the heater to heat the substrate. This benefit can be enhanced by utilising complimentary metal-selenide-semiconductor structures to support the interdigitated electrode structures [47].

3. Gas Sensing Parameters

A gas sensor is a device that detects the presence or concentration of gases in the atmosphere. By altering the material's resistance inside the sensor, the sensor generates a corresponding potential difference based on the gas concentration, which can be measured as output voltage. The type and concentration of the gas can be derived from this voltage value. Gas sensors are available to identify oxygen, carbon dioxide, nitrogen, and methane gases. They are also frequently used in gadgets that check the air quality in workplaces and industries (look for harmful gas leaks). Portable and fixed gas sensors come in various sizes, sensing capabilities, and ranges. They typically connect to an interface or audible alarm and are frequently a component of a more extensive embedded system of security systems. The most common parameters to evaluate gas-sensing performance are optimal working temperature, stability, sensitivity, repeatability, selectivity, the lowest detection limits, response, and recovery time (shown in Figure 3).

- **Limit of detection (LOD):** The lowest gas concentration, or minimum detection concentration, at which the sensor can respond under specific circumstances.

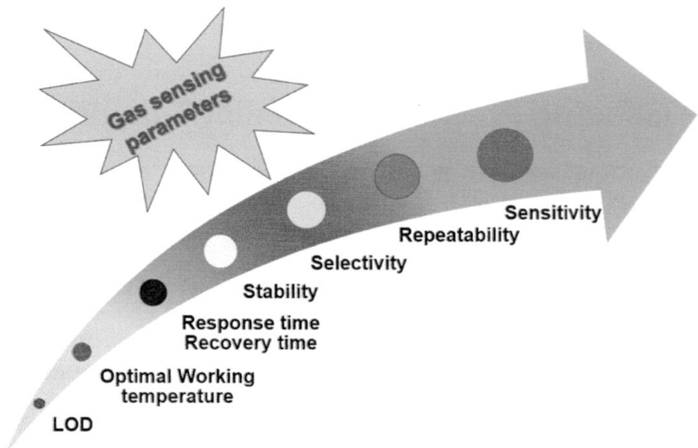

Figure 3. Schematic diagram shows all respective sensing parameters.

- **Response and recovery time:** The response time and recovery time are terms used to describe how long it takes a device to increase its resistance to 10–90% of its final value and how long it takes to decrease the resistance to 90–10% of its final value.
- **Optimal working temperature:** The working temperature affects the sensing properties of most semiconductor-based gas sensors. The sensitive materials can fully stimulate the chemical activity that pushes the gas sensor to its maximum response at the optimal operating temperature. Most semiconductor sensors have high operating temperatures (in 100s of degrees centigrade), resulting in significant power consumption. One of the current challenges is creating a particular gas sensor with a low operating temperature [48].
- **Stability:** An essential characteristic for assessing a sensor is its stability. Whether the sensor can continue operating for a prolonged time while maintaining a similar trend as the initial performance within the predetermined working range, is a concern for the users [49].
- **Selectivity:** Selectivity is a gas sensor's capacity to precisely identify the measured gas in a mixed gas environment.
- **Repeatability:** After measuring the target gas, a semiconductor sensor's resistance should be the same even after repeated use. Thus, repeatability is the ability of a sensor to return to its initial resistance and maintain the sensing performance [50].
- **Sensitivity:** A gas sensor's responsiveness to the target gas is indicated by its sensitivity, S. It depicts the relationship between the gas sensor's electrical characteristics and the target gas concentration. Undoubtedly, the performance of the gas-sensing materials improves with increasing S values. $S = ((R_a - R_g)/R_a)\times 100$ (*n*-type semiconductor) or $S = ((R_g - R_a)/R_a)\times 100$ (*p*-type semiconductor) are typical formulas for expressing the sensitivity. The sensor's

resistance in the presence of air and the resistance in the presence of target gas, respectively, are represented by R_a and R_g [51].

4. Gas Sensing Properties of Metal Selenides

Research has demonstrated that in these materials, called selenides, the specific recognition of gases is achieved by a variety of interactions, such as hydrogen bonding between the structure's surface and gases, coordination of the gas molecules to the central metal ion, and van der Waals interactions of the structure's surface with gases [52,53]. The ability to tune pore size, control the interaction with guest molecules by altering the organic linker molecules' nature and the coordination environment of the constituent metal ions, chemical functionality, and post-synthetic modifications are among the unique properties of selenides that make these materials appealing for gas sensing using various transduction principles. However, these properties also make selenides useful for gas detection. Since very porous materials like selenides effectively concentrate analyte molecules at higher levels than are present in the external atmosphere, they should be intrinsically sensitive for gas or vapour detection utilising mass-sensitive sensors. For instance, Priyanka Chaudhary et al. (2022) reported on the simple, inexpensive, and phosphine-free synthesis of indium selenide (InSe) quantum dots (QDs) based gas sensor using a one-pot synthesis method. With response and recovery times estimated at 7.5 and 6.3 s, respectively, the sensor demonstrated an excellent response of 4.41 for LPG gas [13]. A new light-modulated bi-functional gas sensor based on 2D InSe nanosheets modified with Au nanoparticles was shown by Lu Zhang et al. When exposed to visible light, the constructed sensor recognised nitrogen dioxide (NO_2) with a reversible and exceptionally high response. There was a ten times rise in sensitivity (1192%) compared to dark conditions respectively [3]. In a different study, Wei Zheng et al. show that indium selenide (InSe) nanosheets work incredibly well for the real-time detection of harmful NO_2. The sensitivity of the sensor is high, about 2.75. When exposed to 10 ppm NO_2 at room temperature (395 nm light), the InSe sensor responds quickly (6 s) and recovers quickly (169 s) [28]. In 2022, S. Liu et al. used a hydrothermal technique to synthesise nanostructured molybdenum selenide ($MoSe_2$) with composited phases. To enhance the gas sensing capabilities of the products, metallic particles were added. At the ideal operating temperature of 1200 °C, the sensitivity of 1T/2H-$MoSe_2$ nanosheets modified by Cu nanoparticles is 17.73 (50 ppm NO_2), and their response and recovery times are 23 and 25 s, respectively [6]. A stable chemiresistive H_2S gas sensor based on liquid exfoliated $MoSe_2$ nanoflakes that operates at a moderate temperature of 200 °C is reported here by Ravindra Kumar Jha et al. in 2019. At around 50 ppb of H_2S concentration, the p-type $MoSe_2$ gas sensor device responded with 18.57%. The recovery/response time is 43/15 [54]. Using bi- and multi-layered 2H version of tungsten di-selenide (2H-WSe_2) nanosheets,

Abderrahim Moumen et al. present a simple and reliable room-temperature NO_2 sensor that exhibits high sensing capabilities 361% for 6 ppm at RT (shown in Figure 4) [2].

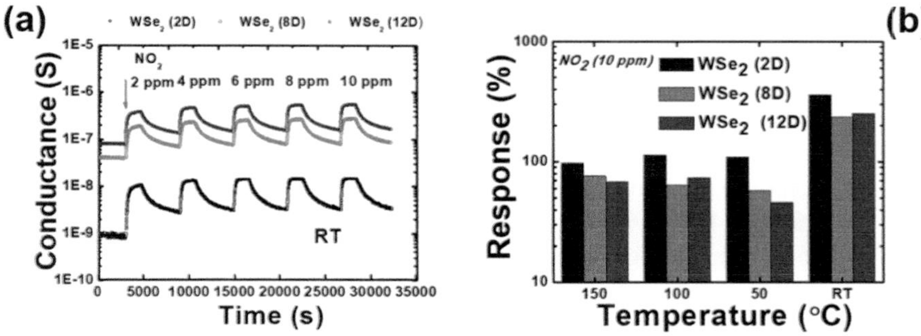

Figure 4. (a) Dynamic response of 2H-WSe_2 (2D), 2H-WSe_2 (8D), and 2H-WSe_2 (12D) sensors at RT and (b) responses of the three sensors toward 10 ppm of NO_2 at different temperatures (RT, 50, 100, and 150 °C) [2].

Tian Wang, Guangfu Liu, et al. produced an ammonia sensor based on a tungsten selenide (SnO_2/WSe_2) nanostructure at ambient temperature in their work in 2022. The SnO_2/WSe_2 composite sensor responded to 5 ppm of ammonia gas with an 87.07% rate, which was 2.58 times greater than that of pristine WSe_2, according to the experimental data. In the meantime, 24 s/40 s was the response/recovery time for the SnO_2/WSe_2 composite sensor [55]. By thermally evaporating $SnSe_2$ powder that was produced in-house, Sanju Rani et al. in 2021 were able to form a thin layer of $SnSe_2$ that was around 150 nm thick. At 30 °C, the SnSe-$SnSe_2$ nanostructured sensor responded with approximately 112% of its sensitivity to 5 ppm NO_2. The periods for response and recovery were approximately 15 and 10 seconds, respectively [56]. In 2019, Professor Leonardo Camargo Moreira et al. use first-principles calculations to examine the gas-sensing capabilities of $SnSe_2$ and use a gas sensor made of few-layer $SnSe_2$ created by chemical vapour deposition to verify predictions. At ambient temperature, the sensitivity is around 10% when exposed to 1 ppm NO_2 [37]. A novel NO_2 gas sensor is presented by Wenlong Yang et al. based on films of tin selenide (SnSe) nanosheet arrays that are formed via vacuum thermal evaporation on a range of substrates. At room temperature, a response rate of 700% and a response time of 12 s were attained at a high NO_2 gas concentration of 10 ppm [34]. Sanju Rani et al. used a thermal evaporation process that is commercially feasible to create a SnSe nanostructured film. For 5 ppm NO_2 at room temperature, the SnSe nanostructured film-based gas sensor responded with around 174%, and its response rate demonstrated a low limit of detection (LOD) of 345 ppb with 106%. It displayed a quick response time of 5 s and a roughly 232 s recovery time [46]. Wei Liu et al. created a unique chemiresistive-type sensor that uses AuPt bimetal-decorated $SnSe_2$ microflowers to detect sub-ppm NO_2. The bimetal precursors were synthesised by hydrothermal treatment and in situ chemical reduction on the microflowers' petals. At 130 °C, the sensor responds well 4.62 to 8 ppm NO_2 gas, with a response time and recovery time of 82/137 [57]. A thick Sn layer was coated on the glass substrate by Lee et al. to create the $SnSe_2$/SnSe heterostructured film. This film-based gas sensor had a response value of 75% for 1 ppm of NO_2 at room temperature [58]. In 2022, Tingtinga Li et al. used a hydrothermal technique to produce a coral-like tin diselenide/MOFs-derived flower-like tin dioxide ($SnSe_2$/SnO_2) nanocomposite film sensor for the detection of NO_2 gas. Specifically, the $SnSe_2$/SnO_2 nanocomposite sensor's response value to 20 ppm NO_2 gas improved noticeably from 7.4 to 12.6 under UV illumination, and the response and recovery durations to 10 ppm NO_2 gas decreased from 110 s/160 s to 80 s/144 s [59]. Sanju Rani et al. observed a temperature-dependent *n-p-n* switching NO_2 gas sensor with high selectivity operating at room temperature (RT) after using the thermal evaporation approach to deposit an *n*-$SnSe_2$/*p*-SnO/*n*-SnSe heterojunction. For 5 ppm NO_2, the device response at RT was 256%. The recovery and response times were 272 and 34 seconds, respectively [60]. Thermal evaporation was used by Sanju et al. to produce $SnSe_2$/SnO_x/SnSe heterojunction thin films with varying thicknesses of 137, 241, and 297 nm. The device's (137 nm thickness) sensitivity at room temperature was 151% for five ppm NO_2. The corresponding response and recovery times were 134 and 494 s [61]. Table 1 provides a summary of the above discussion.

5. Gas Sensing Mechanism

The metal selenide sensor's sensing mechanism relies on the chemical mechanism and charge transfer between the adsorbed gases and the sensing devices. The fundamental idea of gas sensing is the way that exposure to a gas alters the electrical conductivity and resistance of the sensing film. Since atmospheric oxygen makes up the majority of air, the interaction between metal chalcogenides and atmospheric oxygen is essential to gas sensing.

Understanding that there are two forms of gases in nature, one of which is reducing, such as CO, H_2, CH_4, NO, NH_3, etc., can help you understand the sensing mechanism of selenium materials [62]. The following reactions show how charges transfer in a reducing gas [26].

$$NH_3\ (gas) \rightarrow NH_3\ (ads) + e^-$$

$$4NH_3^+(ads) + 30^+\ (ads) \rightarrow 2N_2 + 6H_2O + 3e^-$$

The other category is oxidising, which includes NO_2, CO_2, O_2, and other gases. The following reactions demonstrate charge transfer in an oxidising gas [63].

$$O_2\ (gas) + e^- \rightarrow O_2^-\ (ads)$$

Table 1. The comparison between all gas sensing parameter for selenide based gas sensor

Material	Structure	Gas	Gas conc. (ppm)	Temp. (°C)	Response	Response time (s)	Recovery time (s)	Ref.
InSe	quantum dots	LPG	2.0 vol%	500	4.41[a]	7.5	6.3	[13]
Au-InSe	nanosheets	NO_2	0.1	RT	1192[b]	~	~	[3]
InSe	nanosheets	NO_2	10	RT	2.75[a]	6	169	[28]
$MoSe_2$	nanoparticle	NO_2	50	1200	17.73[c]	23	25	[6]
$MoSe_2$	nanoflakes	H_2S	50 ppb	200	18.57[b]	43	15	[54]
WSe_2	nanosheets	NO_2	6	RT	361[c]	~	~	[2]
SnO_2/WSe_2	nanostructure	NH_3	5	RT	87.07[b]	24	40	[55]
$SnSe_2$	nanostructure	NO_2	5	RT	112[d]	15	10	[56]
$SnSe_2$	Thin film	NO_2	1	RT	10[b]	~	~	[37]
SnSe	nanosheets	NO_2	10	RT	700[c]	12	~	[34]
SnSe	nanostructure	NO_2	5	RT	174[e]	5	232	[46]
$AuPt/SnSe_2$	micro flower	NO_2	8	130	4.62[a]	82	137	[57]
$SnSe_2/SnSe$	heterostructure	NO_2	1	RT	75[f]	~	~	[58]
$SnSe_2/SnO_2$	nanocomposite	NO_2	20	RT	7.4[a]	110	160	[59]
$SnSe_2/SnO/SnSe$	heterojunction	NO_2	5	RT	256[b]	34	272	[60]
$SnSe_2/SnO/SnSe$	heterojunction	NO_2	5	RT	151[b]	134	494	[61]

a. $R = R_g/R_a$
b. $R = [(R_g - R_a)/R_a] \times 100$,
c. $R = [(R_a - R_g)/R_a] \times 100$,
d. $R = (R_g/R_a) \times 100$
e. $R = (R_a/R_g) \times 100$
f. $S = [(I_g - I_0)/I_0] \times 100$

$$NO_2 \text{ (gas)} + e^- \rightarrow NO_2^- \text{ (ads)}$$

$$NO_2 \text{ (gas)} + O_2^- \text{ (ads)} \rightarrow NO_2^-\text{(ads)} + 2O_2^- \text{ (ads)}$$

On the surface of the sensor, atmospheric oxygen molecules are first ionosorbed to create oxygen ionic species (O_2^- and O^-). From the conduction band of *n*-type sensing materials, newly created oxygen species take electrons. This causes the depletion layer to form at the grain boundaries together with the Schottky barrier. In the end, this causes the sensing film's resistance to increase. More holes are transferred from the gas to the material, when it is exposed to oxidizing gases. Electrons go from the sensor material to the gas because oxidizing gases have the ability to bound the electrons. Therefore, the overall sensor resistance more increases (shown in figure 5) [64].

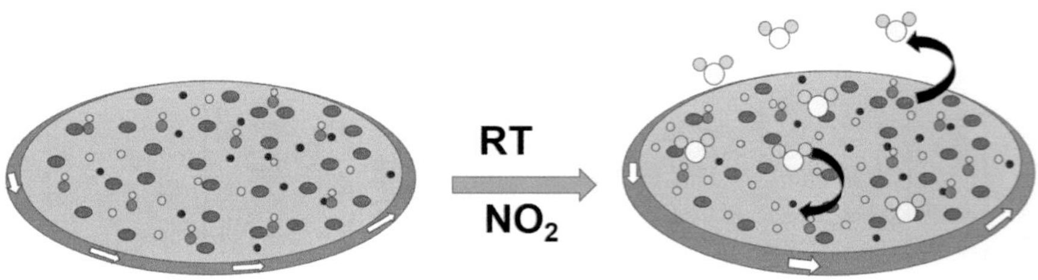

Figure 5. Schematic diagram of gas sensing mechanism at room temperature for oxidizing gas (NO_2).

6. Conclusion and Future Prospective

The current advancements in metal selenides-based gas sensors are outlined in this chapter. Much work has gone into creating high-performing sensing devices using nanostructure, which has a high surface to volume ratio. Because certain results were so sensitive and selective at ambient temperature, they demonstrated amazing sensing performance. These results provide enormous promise for practical devices. The principles of creating practical gadgets continue to provide difficulties, regardless of advancements. The limit of detection is one of the main problems. The other is that these sensors are expensive when used in real-world systems and have low response times and delayed recoveries. In addition, further developments in the selenide-based gas sensor's cost management, quick response and recovery time, and response enhancement are anticipated. Therefore, by using high-performing metal selenide-based materials to create the heterostructure and composite (p-n junction), we may decrease response time, recovery time, and production cost while simultaneously improving response.

References

1. S.-K. Liu, S. Cai, Y. Chen, B. Xiao, P. Chen, X.-D. Xiang, The effect of pollutional haze on pulmonary function, J. Thorac. Dis., 8 (2016) E41.
2. A. Moumen, R. Konar, D. Zappa, E. Teblum, I. Perelshtein, R. Lavi, S. Ruthstein, G.D. Nessim, E. Comini, Robust room-temperature NO_2 sensors from exfoliated 2D few-layered CVD-grown bulk tungsten di-selenide (2H-WSe_2), ACS Appl. Mater. Interfaces, 13 (2021) 4316-4329.
3. L. Zhang, Z. Li, J. Yang, J. Zhou, Y. Zhang, H. Zhang, Y. Li, A fully integrated flexible tunable chemical sensor based on gold-modified indium selenide nanosheets, ACS Sensors, 7 (2022) 1183-1193.
4. S. Hong, M. Wu, Y. Hong, Y. Jeong, G. Jung, W. Shin, J. Park, D. Kim, D. Jang, J.-H. Lee, FET-type gas sensors: A review, Sensors Actuators B Chem. 330 (2021) 129240.
5. J. Dai, O. Ogbeide, N. Macadam, Q. Sun, W. Yu, Y. Li, B.-L. Su, T. Hasan, X. Huang, W. Huang, Printed gas sensors, Chem. Soc. Rev., 49 (2020) 1756-1789.
6. S. Liu, Y. Liu, H. Li, Y. Bai, J. Xue, R. Xu, M. Zhang, G. Chen, Performance enhancement of gas sensing by modification of molybdenum selenide nanosheets with metal nanoparticles, Nanotechnology, 33 (2022) 215501.
7. R. Guo, Y. Han, C. Su, X. Chen, M. Zeng, N. Hu, Y. Su, Z. Zhou, H. Wei, Z. Yang, Ultrasensitive room temperature NO_2 sensors based on liquid phase exfoliated WSe_2 nanosheets, Sensors Actuators B Chem., 300 (2019) 127013.
8. B. Urasinska-Wojcik, T.A. Vincent, M.F. Chowdhury, J.W. Gardner, Ultrasensitive WO_3 gas sensors for NO_2 detection in air and low oxygen environment, Sensors Actuators B Chem. 239 (2017) 1051-1059.
9. N. Mkhize, K. Murugappan, M.R. Castell, H. Bhaskaran, Electrohydrodynamic jet printed conducting polymer for enhanced chemiresistive gas sensors, J. Mater. Chem. C, 9 (2021) 4591-4596.
10. M.-S. Yao, W.-H. Li, G. Xu, Metal–organic frameworks and their derivatives for electrically-transduced gas sensors, Coord. Chem. Rev., 426 (2021) 213479.
11. Y. Kim, K.C. Kwon, S. Kang, C. Kim, T.H. Kim, S.-P. Hong, S.Y. Park, J.M. Suh, M.-J. Choi, S. Han, Two-dimensional NbS_2 gas sensors for selective and reversible NO_2 detection at room temperature, ACS Sensors, 4 (2019) 2395-2402.
12. E. Singh, M. Meyyappan, H.S. Nalwa, Flexible graphene-based wearable gas and chemical sensors, ACS Appl. Mater. Interfaces, 9 (2017) 34544-34586.
13. P. Chaudhary, A. Singh, S. Sikarwar, B.C. Yadav, One-pot synthesis of phosphine-free indium selenide (InSe) QDs and their structural characterization for LPG and humidity sensing, J. Mater. Sci. Mater. Electron., 33 (2022) 11802-11812.
14. H.H. Afify, I.K. Battisha, Oxygen interaction with CdS based gas sensors by varying different preparation parameters, J. Mater. Sci. Mater. Electron., 11 (2000) 373-377.
15. M. Donarelli, S. Prezioso, F. Perrozzi, F. Bisti, M. Nardone, L. Giancaterini, C. Cantalini, L. Ottaviano, Response to NO_2 and other gases of resistive chemically exfoliated MoS_2-based gas sensors, Sensors Actuators B Chem., 207 (2015) 602-613.
16. D. Liu, Z. Tang, Z. Zhang, Comparative study on NO_2 and H_2S sensing mechanisms of gas sensors based on WS2 nanosheets, Sensors Actuators B Chem., 303 (2020) 127114.
17. M. Shafiei, J. Bradford, H. Khan, C. Piloto, W. Wlodarski, Y. Li, N. Motta, Low-operating temperature NO_2 gas sensors based on hybrid two-dimensional SnS2-reduced graphene oxide, Appl. Surf. Sci., 462 (2018) 330-336.

18. L. Liu, Q. Yang, Z. Wang, H. Ye, X. Chen, X. Fan, G. Zhang, High selective gas detection for small molecules based on germanium selenide monolayer, Appl. Surf. Sci., 433 (2018) 575-581.
19. S. Bandyopadhyay, B. Chatterjee, P. Nag, A. Bandyopadhyay, Nanocrystalline PbS as ammonia gas sensor: Synthesis and characterization, CLEAN – Soil, Air, Water. 43 (2015) 1121-1127.
20. Y. Wang, C. Tang, M. Su, Y. Ji, L. Xie, Q. Yang, A. Du, Y. Zhou, J. Yang, Highly sensitive, humidity-tolerant and flexible NO_2 sensors based on nanoplate Bi_2Se_3 film, Chinese Chem. Lett., 34 (2023) 107981.
21. D.J. Late, T. Doneux, M. Bougouma, Single-layer $MoSe_2$ based NH_3 gas sensor, Appl. Phys. Lett., 105 (2014) 233103.
22. H. Zhang, Q. Pan, Y. Zhang, Y. Zhang, D. Zhang, High-performance sulfur dioxide gas sensor based on graphite-phase carbon-nitride-functionalized tin diselenide nanorods composite, Chemosensors, 10 (2022) 401.
23. R. Singh, A.K. Shrivastava, A.K. Bajpai, CdSe reinforced polyaniline nanocomposites as superior material for future applications as gas sensor and diodes, Mater. Res. Express., 6 (2020) 1250a9.
24. H. Ye, L. Liu, Y. Xu, L. Wang, X. Chen, K. Zhang, Y. Liu, S.W. Koh, G. Zhang, SnSe monolayer: A promising candidate of SO_2 sensor with high adsorption quantity, Appl. Surf. Sci., 484 (2019) 33-38.
25. S. Park, S. Kim, W.I. Lee, K.-K. Kim, C. Lee, Room temperature, ppb-level NO_2 gas sensing of multiple-networked ZnSe nanowire sensors under UV illumination, Beilstein J. Nanotechnol., 5 (2014) 1836-1841.
26. M. Mishra, A. Dash, A. Sharma, M. Khanuja, G. Gupta, CO sensing properties of nanostructured WSe_2/GaN and $MoSe_2$/GaN based gas sensors, Phys. E Low-Dimensional Syst. Nanostructures, 139 (2022) 115147.
27. M. Li, F. Yang, Y. Hsiao, C. Lin, H. Wu, S. Yang, H. Li, C. Lien, C. Ho, H. Liu, Low-voltage operational, low-power consuming, and high sensitive tactile switch based on 2D layered InSe tribotronics, Adv. Funct. Mater., 29 (2019) 1809119.
28. W. Zheng, C. Yang, Z. Li, J. Xie, C. Lou, G. Lei, X. Liu, J. Zhang, Indium selenide nanosheets for photoelectrical NO_2 sensor with ultra sensitivity and full recovery at room temperature, Sensors Actuators B Chem., 329 (2021) 129127.
29. L. Zhang, Z. Li, J. Liu, Z. Peng, J. Zhou, H. Zhang, Y. Li, Optoelectronic gas sensor based on few-layered InSe nanosheets for NO_2 detection with ultrahigh antihumidity ability, Anal. Chem., 92 (2020) 11277-11287.
30. R. Wen, A. Wei, L. Tao, D. Luo, J. Liu, Y. Yang, Z. Xiao, Z. Liu, Y. Zhao, Hydrothermal synthesis of WSe_2 films and their application in high-performance photodetectors, Appl. Phys. A., 124 (2018) 1-8.
31. W. Shi, M. Gao, J. Wei, J. Gao, C. Fan, E. Ashalley, H. Li, Z. Wang, Tin selenide (SnSe): Growth, properties, and applications, Adv. Sci., 5 (2018) 1700602.
32. L. Li, Y. Yu, G.J. Ye, Q. Ge, X. Ou, H. Wu, D. Feng, X.H. Chen, Y. Zhang, Black phosphorus field-effect transistors, Nat. Nanotechnol., 9 (2014) 372-377.
33. J. Wang, G.F. Yang, J.J. Xue, J.M. Lei, D.J. Chen, H. Lu, R. Zhang, Y.D. Zheng, A reusable and high sensitivity nitrogen dioxide sensor based on monolayer SnSe, IEEE Electron Device Lett., 39 (2018) 599-602. doi: 10.1109/LED.2018.2806367.
34. W. Yang, C. Zhao, B. Du, R. Wu, X. Lai, Y. He, J. Jian, SnSe nanosheet arrays film for trace NO_2 detection at room temperature, Sensors Actuators B Chem., 370 (2022) 132407.

35. P. Manou, J.A. Kalomiros, A.N. Anagnostopoulos, K. Kamba, Optical properties of $SnSe_2$ single crystals, Mater. Res. Bull., 31 (1996) 1407.
36. E.P. Mukhokosi, S.B. Krupanidhi, K.K. Nanda, Band gap engineering of hexagonal $SnSe_2$ nanostructured thin films for infra-red photodetection, Sci. Rep., 7 (2017) 1-10. https://doi.org/10.1038/s41598-017-15519-x.
37. Ó.L.C. Moreira, W.Y. Cheng, H.R. Fuh, W.C. Chien, W. Yan, H. Fei, H. Xu, D. Zhang, Y. Chen, Y. Zhao, Y. Lv, G. Wu, C. Lv, S.K. Arora, C.Ó. Coileáin, C. Heng, C.R. Chang, H.C. Wu, High selectivity gas sensing and charge transfer of $SnSe_2$, ACS Sensors, 4 (2019) 2546-2552. https://doi.org/10.1021/acssensors.9b01461.
38. M. Chen, Z. Li, W. Li, C. Shan, W. Li, K. Li, G. Gu, Y. Feng, G. Zhong, L. Wei, C. Yang, Large-scale synthesis of single-crystalline self-standing $SnSe_2$ nanoplate arrays for wearable gas sensors, Nanotechnology, 29 (2018).
39. M.A. Airo, S. Gqoba, M.P. Kalenga, S. Govindraju, M.J. Moloto, N. Moloto, Synthesis and characterization of indium monoselenide nanosheets: A proposed pseudo top-down mechanism, J. Cryst. Growth, 406 (2014) 1-7.
40. S. Jamil, F. Farooq, S.R. Khan, M.R.S.A. Janjua, Synthesis of WSe_2 nanorods by selenium powder precursor for photocatalytic application and fuel additive, J. Clust. Sci., 32 (2021) 1061-1073.
41. X. Zhang, M. Xue, X. Yang, G. Luo, F. Yang, Hydrothermal synthesis and tribological properties of $MoSe_2$ nanoflowers, Micro Nano Lett., 10 (2015) 339-342.
42. R.K. Rai, S. Islam, A. Roy, G. Agrawal, A.K. Singh, A. Ghosh, N. Ravishankar, Morphology controlled synthesis of low bandgap $SnSe_2$ with high photodetectivity, Nanoscale, 11 (2019) 870-877.
43. M. Kumar, S. Rani, P. Vashistha, A. Pandey, G. Gupta, S. Husale, V.N. Singh, Low bias operated, fast response SnSe thin film Vis-NIR photodetector on glass substrate using one-step thermal evaporation technique, J. Alloys Compd. 879 (2021) 160370. https://doi.org/10.1016/j.jallcom.2021.160370.
44. D. Zheng, J. Shiogai, K. Fujiwara, A. Tsukazaki, Pulsed-laser deposition of InSe thin films for the detection of thickness-dependent bandgap modification, Appl. Phys. Lett., 113 (2018).
45. X. Wang, Y. Gong, G. Shi, W.L. Chow, K. Keyshar, G. Ye, R. Vajtai, J. Lou, Z. Liu, E. Ringe, Chemical vapor deposition growth of crystalline monolayer $MoSe_2$, ACS Nano, 8 (2014) 5125-5131.
46. S. Rani, M. Kumar, H. Sheoran, R. Singh, V.N. Singh, Rapidly responding room temperature NO_2 gas sensor based on SnSe nanostructured film, Mater. Today Commun., 30 (2022) 103135. https://doi.org/10.1016/j.mtcomm.2022.103135.
47. M.V. Nikolic, V. Milovanovic, Z.Z. Vasiljevic, Z. Stamenkovic, Semiconductor gas sensors: Materials, technology, design, and application, Sensors, 20 (2020) 6694.
48. X. Xu, S. Ma, X. Xu, S. Pei, T. Han, W. Liu, Transformation synthesis of heterostructured SnS_2/ZnS microspheres for ultrafast triethylamine detection, J. Alloys Compd., 868 (2021) 159286.
49. L. Meng, W. Bu, Y. Li, Q. Qin, Z. Zhou, C. Hu, X. Chuai, C. Wang, P. Sun, G. Lu, Highly selective triethylamine sensing based on SnO/SnO_2 nanocomposite synthesized by one-step solvothermal process and sintering, Sensors Actuators B Chem., 342 (2021) 130018.
50. F. Meng, Z. Liao, C. Xing, Z. Yuan, R. Zhang, H. Zhu, J. Li, Preparation of SnO_2/SiO_2 nanocomposites by sol-gel method for enhancing the gas sensing performance to triethylamine, J. Alloys Compd., 893 (2022) 162189.

51. X. Liang, J. Zhang, K. Zhang, X. Yang, M. Zhang, The modification effect of Fe_2O_3 nanoparticles on ZnO nanorods improves the adsorption and detection capabilities of TEA, Inorg. Chem. Front., 9 (2022) 259-266.
52. R.J. Kuppler, D.J. Timmons, Q.-R. Fang, J.-R. Li, T.A. Makal, M.D. Young, D. Yuan, D. Zhao, W. Zhuang, H.-C. Zhou, Potential applications of metal-organic frameworks, Coord. Chem. Rev., 253 (2009) 3042-3066.
53. L.R. MacGillivray, Metal-Organic Frameworks: Design and Application. John Wiley & Sons (2010).
54. R.K. Jha, J.V. D'Costa, N. Sakhuja, N. Bhat, $MoSe_2$ nanoflakes based chemiresistive sensors for ppb-level hydrogen sulfide gas detection, Sensors Actuators B Chem. 297 (2019) 126687.
55. T. Wang, G. Liu, D. Zhang, D. Wang, F. Chen, J. Guo, Fabrication and properties of room temperature ammonia gas sensor based on SnO_2 modified WSe_2 nanosheets heterojunctions, Appl. Surf. Sci., 597 (2022) 153564.
56. S. Rani, M. Kumar, Y. Singh, M. Tomar, A. Sharma, V. Gupta, V.N. Singh, NO_2 gas sensor based on SnSe/$SnSe_2$ pn Hetrojunction, J. Nanosci. Nanotechnol., 21 (2021) 4779-4785. https://doi.org/10.1166/jnn.2021.19278.
57. W. Liu, D. Gu, X. Li, AuPt bimetal-functionalized $SnSe_2$ microflower-based sensors for detecting sub-ppm NO_2 at low temperatures, ACS Appl. Mater. Interfaces, 13 (2021) 20336-20348.
58. L. Lee, C.W. Chen, A. Manikandan, S.H. Lee, Z.M. Wang, Y.L. Chueh, Phase-engineered $SnSe_x$ toward $SnSe_2$/SnSe heterostructure with improved thermal conductance by a low-temperature plasma-assisted chemical vapor reaction, Nano Energy, 44 (2018) 419-429.
59. T. Li, D. Zhang, Q. Pan, M. Tang, S. Yu, UV enhanced NO_2 gas sensing at room temperature based on coral-like tin diselenide/MOFs-derived nanoflower-like tin dioxide heteronanostructures, Sensors Actuators B Chem., 355 (2022) 131049. https://doi.org/10.1016/j.snb.2021.131049.
60. S. Rani, M. Kumar, P. Garg, R. Parmar, A. Kumar, Y. Singh, V. Baloria, U. Deshpande, V.N. Singh, Temperature-dependent n–p–n switching and highly selective room-temperature n-$SnSe_2$/p-SnO/n-SnSe heterojunction-based NO_2 gas sensor, ACS Appl. Mater. Interfaces, 14 (2022) 15381-15390. https://doi.org/10.1021/acsami.1c24679.
61. S. Rani, M. Kumar, P. Garg, R. Yadav, Y. Singh, A. Kumar, B. Govind, U. Deshpande, S. Hausale, V.N. Singh, Thickness dependent pn switching in $SnSe_2$/SnO_x/SnSe heterojunction-based NO_2 gas sensor as well as photodetector, J. Sci. Adv. Mater. Devices, 8 (2023) 100583.
62. Q. Yue, Z. Shao, S. Chang, J. Li, Adsorption of gas molecules on monolayer MoS_2 and effect of applied electric field, Nanoscale Res. Lett., 8 (2013) 1-7.
63. L. Dong, P. Zheng, Y. Yang, M. Zhang, Z. Xue, Z. Wang, G. Liu, P. Li, K.S. Teh, Y. Su, NO_2 gas sensor based on graphene decorated with Ge quantum dots, Nanotechnology, 30 (2018) 74004.
64. A. Sharma, G. Gupta, Recent development and prospects for metal selenide-based gas sensors, Mater. Sci. Eng. B, 290 (2023) 116333.

CHAPTER
7

Emerging 2D Materials for Gas Sensor

Biswajit Das[1,2] **and Ranajit Ghosh**[1,2*]

[1] Coatings & Surface Engineering Group, CSIR-Central Mechanical Engineering Research Institute, Durgapur 713209, West Bengal, India

[2] Academy of Scientific and Innovative Research (AcSIR), Ghaziabad 201002, Uttar Pradesh, India

1. Introduction

The ever-increasing population and materialistic demand of mankind has resulted in rising environmental degradation like air, water and soil pollution. Hence, the requirement of superior quality gas sensors is becoming an inevitable for especially to detect hazardous gases that cause air pollution. Moreover, the non-invasive techniques of early stage detection of various diseases by detecting volatile organic compounds (VOCs) in the exhaled breath of affected patients is gaining popularity which also requires ultra-sensitive gas sensors. The most popular kind of gas sensor among various others (i.e. optical, calorimetric, SAW type, electrochemical, capacitance based) is a semiconductor based chemiresistive type, which works on the principle of change of conductivity in the presence of gaseous environment. Specifically, in the presence of a gaseous environment, the gas molecules are adsorbed on the sensing surface and consecutive charge transfer occurs between them, which leads to changes in the surface conductivity of the material. So, there are two significant factors of sensing performance: (i) interaction of gas molecules on the surface which largely depends on the catalytic property of the sensing material and it is very sensitive to temperature, (ii) specific surface area of the sensing material since gas sensing is a surface phenomenon. The chemiresistive type of gas sensor was first discovered by

*Corresponding author: r_ghosh@cmeri.res.in

Seiyama et al. using ZnO as semiconductor with a working temperature above 400 °C [1]. Historically, numerous works on chemiresistive gas sensors were carried out thereafter using semiconducting metal oxides as sensing elements due to its chemical stability, abundance, low cost, easy processing etc. For example, Hiroyuki et al. has demonstrated In_2O_3 based carbon monoxide detection at 300 °C [2]. Whereas, Eiichi et al. has developed a SrO loaded SnO_2 based sensor operated at an elevated temperature of 500 °C for the detection of N_2O [3]. In another report, Jiaqiang et al. performed a comparative study on the ethanol (C_2H_5OH) sensing performance of pure and Ru, Mg, Pd, Y, La, V and Na doped ZnO in the operating temperature range of 180–390 °C. They found that ZnO based sensors exhibit excellent sensing performance towards ethanol which further increases with Ru doping. The sensing mechanism was explained on the basis of chemisorption of oxygen ion species where Ru dopant acts as sensitizer [4]. In addition to that, there are several other reports on metal oxide based chemiresistive sensors [5–10] and the basic reason for exhibiting the sensing phenomena of all the metal oxide based sensors is chemisorption. Now, one of the major drawbacks of these metal oxide based sensors is its high power consumption due to elevated operating temperature. The adopted sensor structure for operating at high temperature i.e. Taguchi type which consists of a hollow ceramic tube on which the sensing layer is coated along with a pair of electrodes at both ends of the tube whereas the heated is inserted inside the hollow region. This kind of sensor structure is a hindrance for gas sensor device miniaturization. In addition, lack of selectivity is another adverse behavior of metal oxide based sensors. To overcome all drawbacks of metal oxide based sensors, researchers were searching for new alternative materials which offer sensing characteristics towards room temperature. After the discovery of graphene i.e. the 2-dimension (2D) allotrope of carbon by Novoselov et al., the scientific community started exploring its potential in diverse fields of application including gas sensing e.g. it was first successfully exploited to exhibit sensing behavior towards analytes like NH_3, CO, NO_2 etc. at room temperature [11]. In another study, Ganhua et al. had synthesized exfoliated graphene oxide sheets for the detection of NO_2 at room temperature [12]. Thus, 2D materials like graphene were soon found to be promising candidates for room temperature gas sensing. This trend of application of 2D graphene has forced the researchers to discover other 2D materials like MoS_2, WS_2, $MoSe_2$, WSe_2 etc. which are basically transition metal dichalcogenides with layered structure in bulk counterpart. The atoms of these materials are basically covalently bonded in a plane to form 2D sheets which are then attached with each other via weak van der Waals force. The structures of various 2D materials are represented in Figure 1. Hexagonal boron nitride (hBN) is another layered material with insulating nature possessing similar lattice structure as of graphene consisting of boron and nitrogen. On the other hand, phosphorene was discovered as a 2D counterpart of black phosphorus. Till date, numerous other 2D materials have been discovered and their gas sensing performance was studied. In all of these cases, the sensors

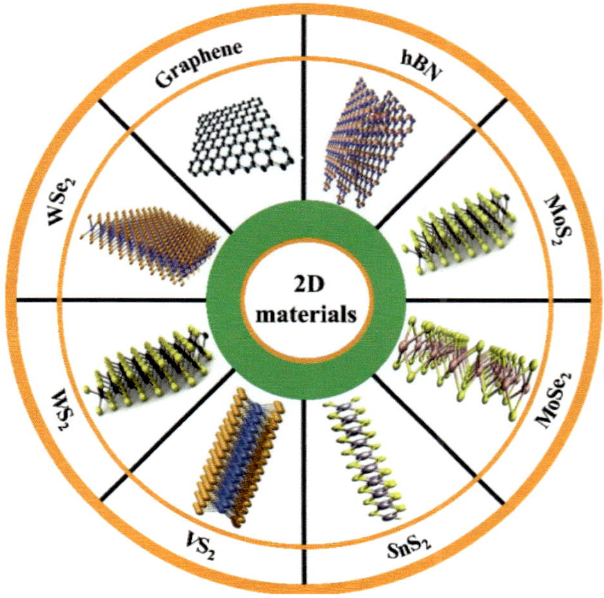

Figure 1. Schematic diagram of structures of various 2D materials.

operate at room temperature and the sensing mechanism relies on physisorption which does not require high operating temperature hence overcomes the drawback of high power consumption and allowing device miniaturization. One of the important properties of 2D materials is their high surface to volume ratio, which is advantageous in many other applications. Since gas sensing is a surface phenomenon, 2D materials as sensing layer avail added benefits of high catalytic activity, better adsorption of gas molecules on the surface and thereby improved sensing.

2. Chemisorption and Physisorption

The consecutive processes during gas sensing in a crude way is (i) adsorption of gas molecules on surface (ii) interaction of gas molecules and then (iii) desorption of molecules from surface. Thus adsorption is the primary process during sensing. Now, adsorptions are of two types and they occur in different temperature regions – chemisorption and physisorption. As the name suggests, chemisorption is the process in which the gas molecules make chemical bonds with the surface which requires elevated temperature [13]. In contrast, physisorption does not involve any chemical interaction of gas molecules with the sensing surface, rather a weak van der Waals interaction occurs during adsorption. Hence, it does not require high temperature for adsorption, in fact the physisorption phenomena decrease with the increasing temperature. In normal condition, oxygen molecules are adsorbed on the sensing surface as different ion species (O_2^-, O^-, O^{2-}) depending on the surface temperature. So, the temperature dependent adsorption phenomenon is represented in Fig. 2.

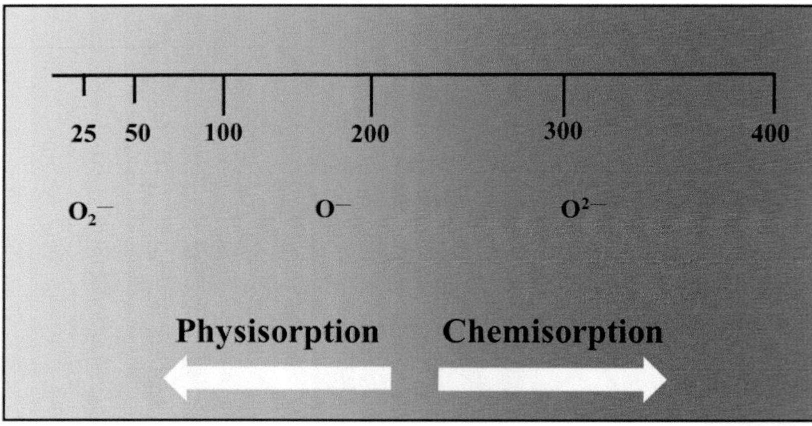

Figure 2. Schematic representation of temperature domain of physisorption and chemisorption and corresponding adsorbed oxygen ion species on the surface of sensing material.

3. 2D Materials in Gas Sensing

The discovery of truly 2D structured graphene and its subsequent application as a room temperature gas sensor attracted immense attention in the research community. Numerous studies have been carried out on its ability to detect various gases at room temperature by modifying its electronic, chemical properties, and which had revolutionized the chemiresistive gas sensing technology than its predecessors (metal oxide based). For instance, Yavari and co-workers have synthesized single and bilayer graphene on copper substrate using CVD method which showed *p*-type behavior to detect both NO_2 and NH_3 at room temperature. They showed that the sensor could detect minute concentration of 100 ppb for NO_2 and 500 ppb for NH_3 having decreasing response time with the increasing gas concentration [14]. In another study done by Alizadeh et al., the composite film of chemically exfoliated graphene and PMMA on IDE was shown to exhibit excellent sensing performance towards formaldehyde at room temperature down towards 50 ppb [15]. However, Pandey et al. found that the sensor was fabricated using a Pd-sputtered graphene oxide sheet to be selective towards hydrogen with a detection limit of 200 ppm. It was observed that the response increases 3 times with the increase of operating temperature between 30 to 75 °C. The effect of Pd on the graphene oxide layer was also investigated to improve the sensor response and Pd layer of 3 nm on 2 nm graphene oxide film was shown to offer the best performance [16]. Till date, graphene is the probably the most studied 2D material in gas sensing application. Due to the overwhelming success of graphene as gas sensing material, research community started to explore other 2D materials which includes transition metal dichalcogenides (TMD), hexagonal boron nitride (hBN), black phosphorous and the very recent discovery of MXenes. Transition metal dichalcogenides are layered materials with a common chemical formula of MX_2, where M stands for transition metals and X is Sulphur (S), selenium (Se) or tellurium (Te). MoS_2 is one of the most studied TMD in a diverse field of application including gas sensing. In a 2011 report published by Topsøe et al., nitric oxide (NO) adsorption properties on MoS_2 surface were examined using IR spectroscopy, STM and DFT studies [17]. It was found that NO is adsorbed very weakly on pristine MoS_2 surface whereas the sulfur vacancy induced surface has enhanced affinity towards NO adsorption. In the next year, Li and co-workers deposited 1–4 layers of mechanically exfoliated MoS_2 nanosheet on Si/SiO_2 substrate to fabricate a FET based sensor to study NO sensing behavior [18]. They found that FET sensors using single layer MoS_2 exhibit dramatic response upon NO exposure at room temperature but the current is unstable passing through the sensor. However, the 2-, 3- and 4-layer MoS_2 sensor exhibit a stable and sensitive response to NO with a detection limit of 0.8 ppm. The AFM image of the synthesized bilayer MoS_2 and the dynamic NO sensing characteristics of the fabricated sensor is represented in Figure 3a,b. In another study, Late et al. had fabricated MoS_2 based FETs based on single and multi-layers and studied their NH_3, NO_2 and humidity sensing properties at room temperature [19]. They compared the sensing performance of the single and multi-layered MoS_2 based FET sensors which interestingly revealed that multi-layered FET sensors exhibit more sensitive and stable performance as compared to single layer MoS_2 based sensor. Latter, Kumar et al. found in their study that the sensing performance of multilayer MoS_2 based sensor can further be improved by photoexcitation of the sensing surface [20]. Briefly, multilayer MoS_2 was grown on Si/SiO_2 substrate and a FET based sensor was fabricated which was employed for NO_2 sensing at room temperature. It was found that response time of the sensor was decreased from ~249 s to ~29 s and the sensitivity was increased upto ~30% upon exposure of the sensing surface to ultraviolet rays. Alongside, there are several other layered TMDs which find application as gas sensor which can be prepared in 2D structure via exfoliation route as well as bottom-up approaches. Typical examples of other TMDs are $MoSe_2$, WS_2, WSe_2, TiS_2, VS_2 etc. Single layer $MoSe_2$ was used by Late et al. to fabricate NH_3 sensor working at room temperature [21]. They have synthesized bulk $MoSe_2$ using chemical vapour transport and exfoliated to single and few layers by mechanical exfoliation on a pre-cleaned Si/SiO_2 substrate. The sensor showed prominent response phenomena towards NH_3 down to 50 ppm which was confirmed by employing Raman Spectroscopy before and after NH_3 exposure. Another investigation by Guo et al. had mechanically exfoliated $MoSe_2$ to few layers large area flakes and fabricated FET based sensor was demonstrated to be responsive towards NH_3 and NO_2 at

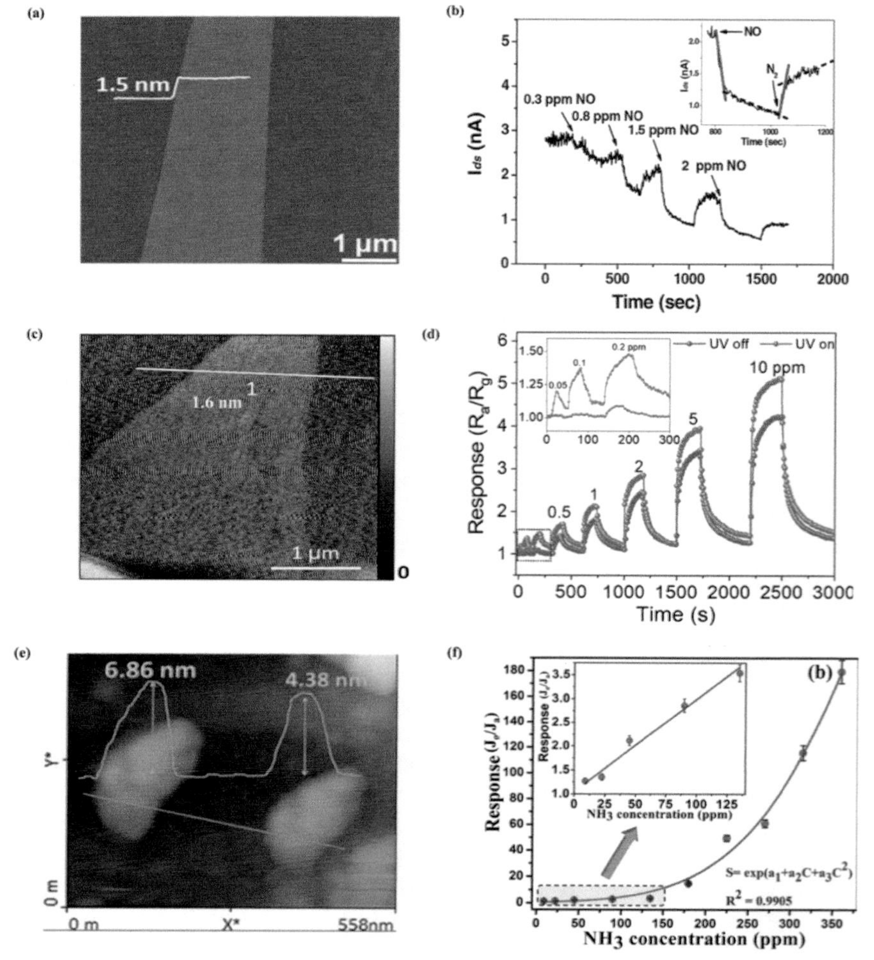

Figure 3. (a) AFM image and (b) dynamic NO sensing characteristics of bilayer MoS_2. (c) AFM image of exfoliated 1.6 nm thick WSe_2 sheet (d) dynamic NO_2 sensing characteristics of the WSe_2 based sensor without and with UV irradiation. (e) AFM image of few layer SnS_2 and (f) dynamic NH_3 sensing curve of the exfoliated SnS_2 based sensor.

room temperature capable of detecting analytes down to 1 ppm [22]. They have integrated the sensor device on human skin and interfaced the sensor wirelessly to the cloud interface. WS_2 also has gained considerable interest in the gas sensing field due to reasons mentioned earlier. For instance, O'Brien and their group synthesised WS_2 film by sulfurizing WO_3 film of varied thicknesses of 2, 5, 10, 20 and 50 nm at temperature below 500 °C [23]. For sensor fabrication, IDE type electrodes were deposited onto the film which could sense NH_3 in a wide range of concentration from 1.4 ppm to 200 ppm at room temperature. In another report, Li et al. had fabricated WS_2 film based sensor by ball milling bulk WS_2 and subsequent dip coating the solution on alumina IDE which showed *p*-type conductivity [24]. The sensor was demonstrated to exhibit almost linear response behaviour towards 1−10 ppm NH_3 with response ranging from ~250% to ~900% at room temperature with excellent selectivity among formaldehyde, ethanol, benzene and acetone. However, the sensor exhibited varied responses to different humidity conditions i.e. the sensor is highly humidity dependent. These shortcomings were addressed by Chen et al. [25]. They demonstrated synthesis of WS_2/PbS composites of different molar ratio for NO_2 detection at room temperature with enhanced anti-humidity ability and full recovery characteristics. They showed that at an optimum molar ratio of WS_2 and PbS (10:1), the sensor exhibited elevated recovery rate from 33.4% to 99.8% as compared to the pure WS_2 based sensor. Moreover, UV irradiation assisted accelerated desorption of water molecules from the sensing surface. The sensor was shown to sense a very low concentration of NO_2 (20 ppb) with descent selectivity and high long term stability. Apart from sulphide based TMD of tungsten, selenide based TMD of tungsten i.e. WSe_2 also found to deliver excellent sensing performance towards various hazardous gases like NO, NO_2 etc. at room temperature. For instance, liquid phase exfoliated WSe_2 in NMP solution was prepared and IDE based sensor was designed by Guo et al. which was found to be selective and sensitive towards NO_2 [26]. They demonstrated that the bulk WSe_2 was exfoliated down to 6 layers and the fabricated sensor possessed very low limit of detection (50 ppb) with a response of 5.06 (R_g/R_a) which is 3 times higher than the sensor fabricated using bulk WSe_2. Laterly, the work done by Yang et al. demonstrates liquid phase exfoliation of bulk WSe_2 into few layers and fabrication of a flexible IDE based sensor for the detection of NO_2 which can detect even minute concentration of 8 ppb at room temperature [27]. They also showed that the sensor exhibits bifunctional selectivity towards NO_2 and trimethylamine i.e. the sensor exhibits highest sensitivity towards NO_2/trimethylamine under UV/without UV illumination with robust flexibility without degrading sensor performance. The AFM image of the exfoliated WSe_2 sheet and the sensing characteristics are shown in Figure 3c,d. In another report, nitric oxide (NO) detection as low as ~15 ppb was reported using 1T/2H 3D-hierarchical WSe_2 nanoscrews derived from WO_x 3D hierarchical nanoscrews through a low-temperature plasma-assisted selenization process by Chen et al. [28]. Briefly, they have deposited 3D nanoscrew structure WO_x on Si/SiO_2 which is then converted to 1T/2H WSe_2 nanoscrew structure by a plasma assisted selenization process. The sensor fabricated using this structure was found to exhibit linear sensing performance with a response of ~28%–322% in the NO concentration range of 0.06–1 ppm at room temperature. They have also shown that the 3D 3D-hierarchical WSe_2 nanoscrew-based sensor exhibits much superior results then the flat WSe_2 sensor. The sensing mechanism was explained on the basis of barrier height modulation of Schottky junction formed at the interface of 1T-metallic and 2H-semiconducting WSe_2. VS_2 is another example of TMD which is less explored in gas sensing field. However, there are some recent reports of hazardous gas sensing properties of exfoliatd VS_2 like NH_3, NO_2 etc. H. Wang et al. had implemented an innovative approach to increase the gas sensing performance of exfoliated VS_2 nanosheets [29]. They have expanded the interlayer spacing between two consecutive VS_2 sheets by solvothermal ammonia intercalation. The inter-layer spacing was modulated to 10 nm, 10.2 nm and 11.2 nm by changing the intercalation parameters and showing that the highest response is delivered by the maximum intercalated VS_2 sheets (11.2 nm) towards NO_2 due to the increment of surface active area. The fabricated sensor exhibited excellent response in a wide range from 150 ppm to the minute concentration of 0.2 ppm with a fast response/recovery time of ~10s/10s and good long term stability. Another report by Kang et al. demonstrates NH_3 detection properties of liquid phase exfoliated VS_2 nanosheets at temperature as low as 40 °C [30]. They could attain the lowest thickness of the exfoliated sheets to 4 nm which was used for sensor fabrication. The sensor exhibited a decent response towards NH_3 from 1 to 10 ppm with a response ranging from 1.17 to 2.57 (R_g/R_a) but the sensor does not completely recover its initial resistance above 4 ppm. The sensing mechanism was explained as the adsorption of NH_3 molecules on the edge sites of VS_2 and subsequent charge transfer from NH_3 to VS_2 surface. Apart from TMDs, there are other metal dichalcogenides possessing layered structure and ease of exfoliating to ultrathin layers. For example, SnS_2 found several applications in sensing different gases. Pyeon et al. fabricated ALD deposited vertically aligned 2D SnS_2 based sensor on flexible Al_2O_3/PI substrate for the detection of NO_2 at room temperature in the low concentration ranges from 100 ppb to 1 ppm with descent response [31]. It is also shown that the sensor exhibits stable sensing performance even after 70000 bending cycles. Das et al. studied the NH_3 sensing performance of exfoliated SnS_2 at room temperature [32]. Figure 3e,f show the AFM image and sensing behaviour of the exfoliated SnS_2 sheet, respectively. Bulk SnS_2 was synthesized using a hydrothermal method which was then exfoliated in liquid phase. The exfoliated SnS_2 solution was used for fabricating Schottky junction based sensors using porous nickel foil as top metal contact. The unique sensor structure and porous nature of the electrode facilitated to attain a high response of 179.4 (I_g/I_a) towards 360 ppm NH_3 with fast response & recovery characteristics. We have also compared the response of the Schottky based sensor with that of SnS_2 based ohmic sensor and exhibited the schottky based sensor to be 140.7 times higher responsive than ohmic based SnS_2 sensor. We have also experimentally confirmed the barrier height modulation at the Schottky junction in the presence of ammonia exposure for improved sensing performance.

4. Emerging 2D Materials for Gas Sensing Application

Till date, numerous 2D materials have been explored in gas sensing field as well as gas sensing technology have been evolved into more accurate, portable, flexible and energy efficient version. However, with the progress of research, experimental and theoretical work on many new 2D materials is going on. Figure 4 shows the typical structure of emerging 2D materials. The most studied material among the newly discovered 2D material series is MXene which was 1st synthesized by Naguib et al. in 2011 [33]. MXenes are generally transition metal nitrides, carbides or carbonitrides with a common chemical formula of "MAX". These are synthesized using the selective etching of "A" layer from MAX phase using harsh chemical agents like HF, NaF, KF, LiF, NH_4HF_2, etc. Typically, "M" are the early transition metals like Ti, Sc, V, Cr, Mo, Nb etc. "X" are generally carbon or nitrogen. Etching the "A" layer gives rise to dangling bonds in "X" or "M" which is compensated by the attachment of various functional groups. Hence, MXenes become functionalized during synthesis. Till date, over 40 different Mxenes have been discovered including $Ti_3C_2T_x$, Ti_2CT_x, Mo_2CT_x, Nb_2CT_x, $Mo_2TiC_2T_x$, $Ti_4N_3T_x$, V_2CT_x, $Ta_4C_3T_x$ and many others. Due to the high intrinsic electrical conductivity of MXenes, it offers high signal to noise ratio (SNR) which is lacking in most of the other 2D materials. The high SNR is very much desirable for the low detection ability of a gas sensor. Among the various discovered MXenes, $Ti_3C_2T_x$ is one of the most widely explored materials in the gas-sensing field [34−39]. For example, Kim et al. had synthesized $Ti_3C_2T_x$ using a chemical etching method from its MAX phase and utilized it to detect different concentrations of various VOCs like ethanol, acetone and ammonia at room

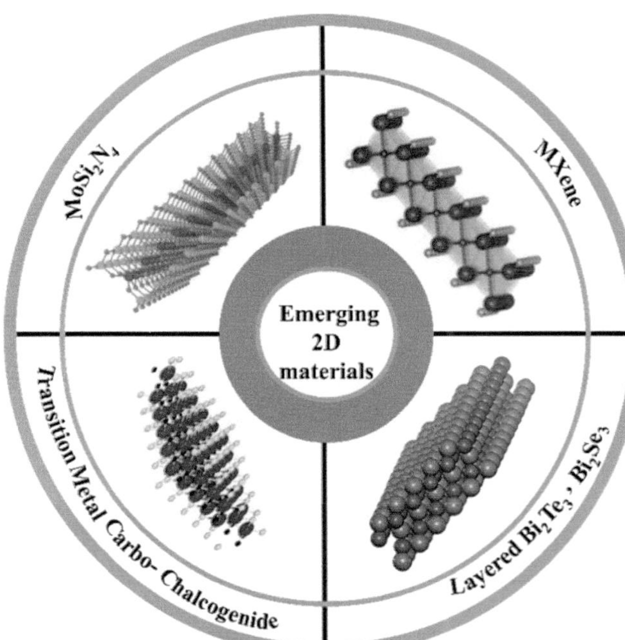

Figure 4. Schematic diagram of structures of various emerging 2D materials.

temperature with descent response [34]. They have shown that the high SNR of $Ti_3C_2T_x$ assisted to attain a lower detection limit down to 50 ppb as compared to other 2D materials. In another study, Shokouh et al. had fabricated $Ti_3C_2T_x$ and PDS-Cl polymer base composite for H_2S detection [35]. They have synthesized composites of different wt% of 4, 6, 8,10, 15, 20 ($Ti_3C_2T_x$: PDS-Cl) by etching and physical blending. Detailed sensing characteristics revealed that the 10 wt% blended sample exhibited the highest response towards H_2S at room temperature with a response around 2 ± 0.2% at 5 ppm and 30-fold increased response at 1 ppm as compared to pristine $Ti_3C_2T_x$. The FESEM image and the H_2S sensing characteristics of 10 wt% sample is shown in Figure 5a,b. Moreover, the sensor was able to detect H_2S concentration as low as 500 ppb. Apart from $Ti_3C_2T_x$, several other MXenes also found application as gas sensors. For example, chemically etched V_2CT_x in combination with PANI was coated on IDE and employed as an ammonia sensor operated at room temperature by Wang and co-workers [40]. The sensor exhibited impressive response of 14.9% at 1 ppm NH_3 and fast response/recovery speed of 9 s/9 s which is better than pristine V_2CT_x based sensor. In addition, the sensor possesses a detection limit as low as 300 ppb with good selectivity and stability. The FESEM image of the composite is exhibited in Figure 5c and the comparative dynamic sensing characteristics of PANI and PANI/V_2CT_x composite is shown in Figure 5d. Another research group led by S. T. Mahmoud synthesized V_2CT_x MXene which was partially converted into V_2O_5 by annealing and employed to detect acetone at room temperature [41]. They observed that the acetone sensing performance of V_2CT_x/V_2O_5 based sensor improved to almost double as compared to pristine V_2CT_x based sensor having response/recovery time of 115 s/180 s. Additionally, it can detect minute acetone concentration down to 250 ppb with a fair response (4.76%) with high selectivity. Despite exhibiting excellent response, very high SNR and ability to detect minute concentration of gas at room temperature, MXene based sensors are lacking in its long term stability needed for practical application due to its chemical instability in atmospheric condition. Extensive research is going on to improve MXene's chemical stability for practical and long term application. In recent years, layered Topological Insulators (TI) are gradually gaining popularity in gas sensing field due to their interesting properties. TI are the materials behaving as an insulator from the interior and as an electrical conductor on the surface. The highly conducting surface is formed due to the gapless Dirac surface state. However, the adsorbed gas molecules on the surface create a narrow gap between the two electronic bands causing change in surface conductivity. This feature is utilized to fabricate gas sensor. Liu et al. had shown that the resistance of solvothermally synthesized layered Bi_2Te_3 (FESEM image in Figure 5e nanoparticles change under ammonia and humid environment (Figure 5f) [42]. In another study, Mousavi et al. had demonstrated physisorption based NO_2 sensing properties of Bi_2Te_3 at room temperature with high selectivity [43]. Recently, we have synthesized different compositions of Bi_2O_3 and thin Bi_2Se_3 nanosheets and employed the material to analyze NH_3 sensing performance

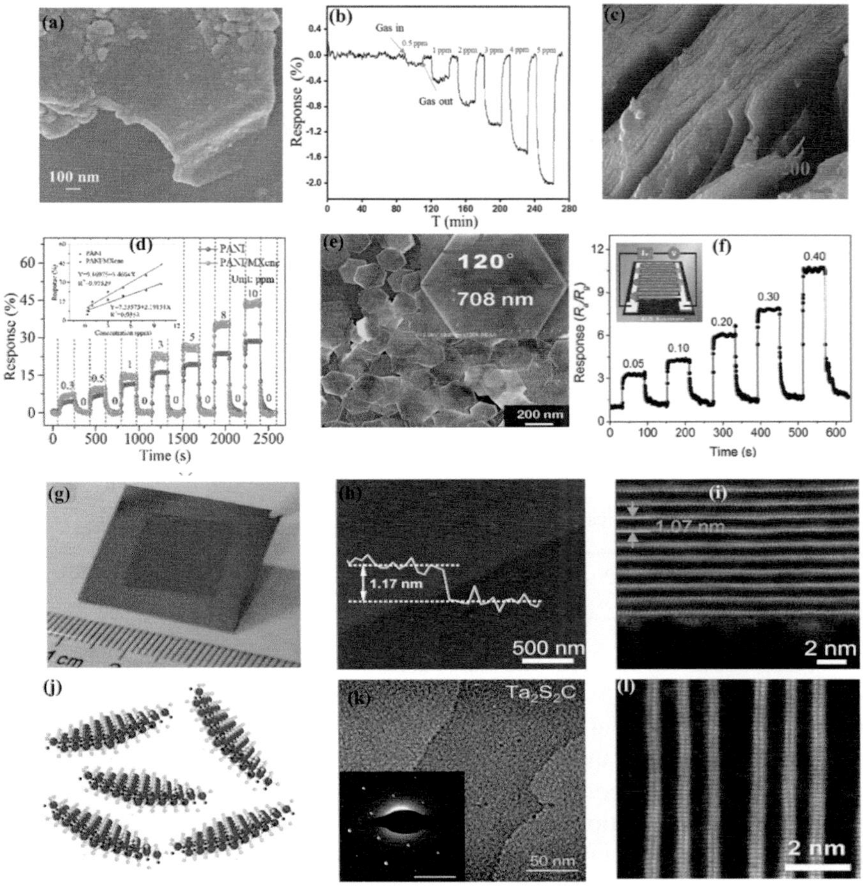

Figure 5. (a) FESEM image and (b) dynamic sensing characteristics of 10 wt% $Ti_3C_2T_x$ based sensor towards H_2S. (c) FESEM image and (d) comparative dynamic sensing characteristics of PANI and PANI/V_2CT_x based sensor towards NH_3. (e) FESEM morphology of hydrothermally synthesized Bi_2Te_3 nanoparticles and (f) corresponding NH_3 sensing characteristics. (g) Optical image and (h) AFM height profile of 2D $MoSi_2N_4$ transferred onto Si/SiO_2 substrate. (i) HAADF-STEM image in the cross-sectional view of thick $MoSi_2N_4$ domain. (j) Schematic diagram of structure of delaminated TMCC layers. (k) TEM image and (l) HAADF-STEM image of single layer Ta_2S_2C.

at room temperature [44]. We found that the main sensing component in the composite is Bi_2Se_3 itself, which shows irreversible sensing nature. Addition of Bi_2O_3 into Bi_2Se_3 in equal amount assists to form *p-n* heterojunction, which is beneficial for optimal sensing performance, and bring the irreversible sensing to reversible. The sensor exhibited a high response of 8.5 towards 180 ppm of NH_3 with a limit of detection of 5 ppm. Moreover, we have experimentally confirmed the physisorption based sensing mechanism with high selectivity and long-term stability over 90 days. Another family of 2D van der Waals (vdW) material was recently discovered by Hong et al. during the synthesis of molybdenum nitride by chemical vapor deposition and introduction of elemental silicon resulting the formation of 2D $MoSi_2N_4$ which shows semiconducting property with a bandgap of 1.94 eV [45]. The centimeter scale 2D layer was transferred to Si/SiO_2 substrate as shown in Fig. 5g. The corresponding AFM image and high-angle annular dark field scanning TEM (HAADF-STEM) image in the crossectional view of thick $MoSi_2N_4$ domain is shown in Figure 5h,i. This discovery led to the theoretical prediction of several other vdW materials with similar crystal structure as $MoSi_2N_4$. These materials can be represented with a general formula of MA_2Z_4, where M is an early transition metal (Mo, W, V, Nb, Ta, Ti, Zr, Hf, or Cr), A is Si or Ge, and Z stands for N, P, or As. They also have predicted 12 different mono-layered materials with MA_2Z_4 structure that are dynamically stable as listed in Table 1.

Table 1: Theoretically predicted dynamically stable 2D vdW materials

$TiSi_2N_4$, $HfSi_2N_4$, $ZrSi_2N_4$, VSi_2N_4, $NbSi_2N_4$, $TaSi_2N_4$, $CrSi_2N_4$, $MoSi_2N_4$, WSi_2N_4, $MoGe_2N_4$, $MoSi_2P_4$, $MoSi_2As_4$

Dong et al. had carried out a DFT analysis on the toxic gas sensing ability of Pd and Ag doped $MoSi_2N_4$ monolayer and found that they are highly sensitive towards CO and NO gas molecules with high binding energy [46]. The recovery speed of NO and CO from the surface of Pd-doped $MoSi_2N_4$ surface was calculated to be very fast (0.03 s and 0.12 s), whereas high temperature is required for fast desorption of these gases from Ag-doped sample. Another study led by Y. Linghu, T. Tong and C. Wu demonstrates DFT analysis of adsorption of various gases like O_2, N_2, CO, NO, NO_2, CO_2, SO_2, H_2O, NH_3 and CH_4 on Cu-doped $MoSi_2N_4$ surface [47]. Their result shows that Cu-doped $MoSi_2N_4$ monolayer can be used as a potential NH_3 sensor at room temperature. Another family of 2D material which is very recently synthesized in 2D form is the Transition Metal Carbo-Chalcogenide (TMCC) by Majed et al. [48]. This new family of 2D material can be considered as the combination of two well-known families of 2D materials: Transition Metal Dichalcogenide (TMD) and MXene. They have synthesized delaminated 2D forms of Ta_2S_2C, Nb_2S_2C from their bulk counterpart using electrochemically assisted exfoliation. The structural schematic of TMCC is drawn in Figure 5j whereas the TEM image and HAADF image of single layer Ta_2S_2C is shown in Figure 5k,l respectively. TMCC can combine the chemical reactivity and catalytic properties of TMDs as well as mechanical robustness and high conductivity of metal carbides. Though, TMCC has not yet found application as gas sensor, it is expected to be used as gas sensor having better sensing performance with lower detection limit than TMDs due to its high conductivity and therefore high SNR.

5. Conclusions

In this chapter, the progression of gas sensing materials used (metal oxides like ZnO, SnO_2, TiO_2, In_2O_3 etc.), sensing technology and the shortcomings of metal oxide based sensors like high operating temperature, lack of selectivity, high power consumption, hindrance for device miniaturization etc. were discussed. We have pointed out the factors responsible for better sensing performance, like catalytic properties, high surface to volume ratio of the sensing material etc. The rise of 2D materials starting from the discovery of graphene to other exfoliated layered materials like MoS_2, $MoSe_2$, WS_2, WSe_2, SnS_2, hBN, VS_2 were illustrated and the advantages brought by these 2D materials in gas sensing technology is cultivated in detail. For example, the 2D materials offer high surface to volume ratio, which is beneficial for higher adsorption of gas molecules. During the synthesis of 2D materials using top down approaches numerous defect sites are created which improves its catalytic property. Moreover, 2D material based sensor relies on physisorption based sensing phenomena, which occur at lower temperature, which helps for device miniaturization and low power consumption. Besides, 2D material based sensor can be fabricated on flexible substrate for wearable applications. All these extraordinary facilities make 2D material based sensor a superior alternative assensing material as compared to metal oxide based sensor. Thereafter, we have described the progression of research of gas sensing performance of different existing 2D materials individually e.g. graphene, MoS_2, $MoSe_2$, WS_2, WSe_2, VS_2, SnS_2 etc. Finally, the current research progress on the emerging 2D materials discovered in recent years has been revealed and elaborately discussed. One of the extensively studied 2D material i.e. MXene ($Ti_3C_2T_x$, V_2CT_x etc.), its synthesis and application and shortcomings has been illustrated with examples. Next, exfoliated layered topological insulators (Bi_2Te_3, Bi_2Se_3 etc.) with exciting electronic properties and their NO_2, humidity and NH_3 sensing properties were drawn. Synthesis of 2D $MoSi_2N_4$ and prediction of 11 other van der Waals structure with similar crystal structure has been emerged as a new family of 2D materials with several theoretical predictions to be used as potential gas sensor. Apart from that, another family of 2D material called TMCC which can be seen as the combination of TMDs and MXenes has been emerged as the latest discovered material in the category of 2D materials though no gas sensing application has been predicted yet using TMCC.

In spite of the significant progress in synthesis and application of emerging 2D materials, there are still several challenges to be overcome. For example, the long term stability for practical of the 2D material based sensors is still not achieved in true manner. Humidity independent sensing is another challenging issue and a headache since 2D material based sensor operates at room temperature. In addition, the fabrication of 2D material based sensor often require high-end instrumental facilities which is a major hindrance for low cost and affordable sensor devices. In future, new 2D materials may be emerged with the technological advancement to surpass the above-mentioned challenges.

References

1. T. Seiyama, A. Kato, K. Fujiishi, M. Nagatani, A new detector for gaseous components using semiconductive thin films determination of benzoic acid in phthalic anhydride by gas liquid chromatography, Anal. Chem., 34 (1962) 1502-1503.
2. N. Yamazoe, N. Miura, K. Moriya, T. Jinkawa, J. Tamaki, H. Yamaura, Indium oxide-based gas sensor for selective detection of CO, Sensors Actuators B Chem., 36 (1996) 325-332.
3. E. Kanazawa, G. Sakai, K. Shimanoe, Y. Kanmura, Y. Teraoka, N. Miura, N. Yamazoe, Metal oxide semiconductor N_2O sensor for medical use, Sensors Actuators, B Chem., 77 (2001) 72-77.
4. J. Xu, J. Han, Y. Zhang, Y. Sun, B. Xie, Studies on alcohol sensing mechanism of ZnO based gas sensors, Sensors Actuators, B Chem., 132 (2008) 334-339.
5. I.D. Kim, A. Rothschild, B.H. Lee, D.Y. Kim, S.M. Jo, H.L. Tuller, Ultrasensitive chemiresistors based on electrospun TiO_2 nanofibers, Nano Lett., 6 (2006) 2009-2013.
6. I.D. Kim, A. Rothschild, D.J. Yang, H.L. Tuller, Macroporous TiO_2 thin film gas sensors obtained using colloidal templates, Sensors Actuators, B Chem., 130 (2008) 9-13.
7. I. Jiménez, J. Arbiol, G. Dezanneau, A. Cornet, J.R. Morante, Crystalline structure, defects and gas sensor response to NO_2 and H_2S of tungsten trioxide nanopowders, Sensors Actuators, B Chem., 93 (2003) 475-485.
8. I. Hotovy, J. Huran, P. Siciliano, S. Capone, L. Spiess, V. Rehacek, Enhancement of H_2 sensing properties of NiO-based thin films with a Pt surface modification, Sensors Actuators, B Chem., 103 (2004) 300-311.
9. I. Raible, M. Burghard, U. Schlecht, A. Yasuda, T. Vossmeyer, V_2O_5 nanofibres: Novel gas sensors with extremely high sensitivity and selectivity to amines, Sensors Actuators, B Chem., 106 (2005) 730-735.
10. H.M. Jeong, H.J. Kim, P. Rai, J.W. Yoon, J.H. Lee, Cr-doped Co_3O_4 nanorods as chemiresistor for ultraselective monitoring of methyl benzene, Sensors Actuators, B Chem., 201 (2014) 482-489.
11. F. Schedin, A.K. Geim, S.V. Morozov, E.W. Hill, P. Blake, M.I. Katsnelson, K.S. Novoselov, Detection of individual gas molecules adsorbed on graphene, Nat. Mater., 6 (2007) 652-655.
12. G. Lu, L.E. Ocola, J. Chen, Gas detection using low-temperature reduced graphene oxide sheets, Appl. Phys. Lett., 94 (2009) 8-11.
13. E. Ciftyurek, Z. Li, K. Schierbaum, Adsorbed oxygen ions and oxygen vacancies: their concentration and distribution in metal oxide chemical sensors and influencing role in sensitivity and sensing mechanisms, Sensors, 23 (2023).
14. F. Yavari, E. Castillo, H. Gullapalli, P.M. Ajayan, N. Koratkar, High sensitivity detection of NO_2 and NH_3 in air using chemical vapor deposition grown graphene, Appl. Phys. Lett., 100 (2012).
15. T. Alizadeh, L.H. Soltani, Graphene/poly (methyl methacrylate) chemiresistor sensor for formaldehyde odor sensing, J. Hazard. Mater., 248-249 (2013) 401-406.
16. P.A. Pandey, N.R. Wilson, J.A. Covington, Pd-doped reduced graphene oxide sensing films for H_2 detection, Sensors Actuators, B Chem., 183 (2013) 478-487.
17. N.Y. Topsøe, A. Tuxen, B. Hinnemann, J.V. Lauritsen, K.G. Knudsen, F. Besenbacher, H. Topsøe, Spectroscopy, microscopy and theoretical study of NO adsorption on MoS_2 and Co-Mo-S hydrotreating catalysts, J. Catal., 279 (2011) 337-351.

18. H. Li, Z. Yin, Q. He, H. Li, X. Huang, G. Lu, D.W.H. Fam, A.I.Y. Tok, Q. Zhang, H. Zhang, Fabrication of single- and multilayer MoS_2 film-based field-effect transistors for sensing NO at room temperature, Small, 8 (2012) 63-67.
19. D.J. Late, Y.K. Huang, B. Liu, J. Acharya, S.N. Shirodkar, J. Luo, A. Yan, D. Charles, U.V. Waghmare, V.P. Dravid, C.N.R. Rao, Sensing behavior of atomically thin-layered MoS_2 transistors, ACS Nano, 7 (2013) 4879-4891.
20. R. Kumar, N. Goel, M. Kumar, UV-activated MoS_2 based fast and reversible NO_2 sensor at room temperature, ACS Sensors, 2 (2017) 1744-1752.
21. D.J. Late, T. Doneux, M. Bougouma, Single-layer $MoSe_2$ based NH_3 gas sensor, Appl. Phys. Lett., 105 (2014).
22. S. Guo, D. Yang, S. Zhang, Q. Dong, B. Li, N. Tran, Z. Li, Y. Xiong, M.E. Zaghloul, Development of a cloud-based epidermal $MoSe_2$ device for hazardous gas sensing, 1900138 (2019) 1-10.
23. M.O. Brien, K. Lee, R. Morrish, N.C. Berner, N. Mcevoy, C.A. Wolden, G.S. Duesberg, Plasma assisted synthesis of WS_2 for gas sensing applications, Chem. Phys. Lett., 615 (2014) 6-10.
24. X. Li, X. Li, Z. Li, J. Wang, J. Zhang, Sensors and Actuators B: Chemical WS_2 nanoflakes based selective ammonia sensors at room temperature, Sensors Actuators B. Chem., 240 (2017) 273-277.
25. X. Chen, J. Hu, P. Chen, M. Yin, F. Meng, Y. Zhang, Chemical UV-light-assisted NO_2 gas sensor based on WS_2/PbS heterostructures with full recoverability and reliable anti-humidity ability, Sensors Actuators B. Chem., 339 (2021) 129902.
26. R. Guo, Y. Han, C. Su, X. Chen, M. Zeng, N. Hu, Y. Su, Z. Zhou, H. Wei, Z. Yang, Ultrasensitive room temperature NO_2 sensors based on liquid phase exfoliated WSe_2 nanosheets, Sensors Actuators, B Chem., 300 (2019).
27. C. Yang, J. Xie, C. Lou, W. Zheng, X. Liu, J. Zhang, Flexible NO_2 sensors based on WSe_2 nanosheets with bifunctional selectivity and superior sensitivity under UV activation, Sensors Actuators, B Chem., 333 (2021) 129571.
28. Y.Z. Chen, S.H. Lee, T.Y. Su, S.C. Wu, P.J. Chen, Y.L. Chueh, Phase-modulated 3D-hierarchical 1T/2H WSe_2 nanoscrews by a plasma-assisted selenization process as high performance NO gas sensors with a ppb-level detection limit, J. Mater. Chem. A, 7 (2019) 22314-22322.
29. H. Wang, Z. Shi, L. Tang, C. Zhu, J. Tang, Q. Wu, W. Song, N. Li, D. Zeng, Tunable and optimal interlayer spacing of 2D VS_2 with high response and fast recovery for NO_2 detection, Sensors Actuators B Chem., 381 (2023) 133445.
30. J. Kang, F. Xu, C. Zhang, F. Li, O.A. Al-Hartomy, A. Al-Ghamdi, S. Wageh, G. Zhao, T. Yang, H. Zhang, Vanadium disulfide nanosheets synthesized by facile liquid-phase exfoliation for ammonia detection with high selectivity, Adv. Electron. Mater., 8 (2022) 2100567.
31. J.J. Pyeon, I.H. Baek, Y.G. Song, G.S. Kim, A.J. Cho, G.Y. Lee, J.H. Han, T.M. Chung, C.S. Hwang, C.Y. Kang, S.K. Kim, Highly sensitive flexible NO_2 sensor composed of vertically aligned 2D SnS_2 operating at room temperature, J. Mater. Chem. C, 8 (2020) 11874-11881.
32. B. Das, S. Behera, B. Satpati, R. Ghosh, Layered SnS_2/porous nickel foil based Schottky junction: An excellent ammonia sensor at room temperature, J. Hazard. Mater., 428 (2022) 128252.
33. M. Naguib, M. Kurtoglu, V. Presser, J. Lu, J. Niu, M. Heon, L. Hultman, Y. Gogotsi, M.W. Barsoum, Two-dimensional nanocrystals produced by exfoliation of Ti_3AlC_2, Adv. Mater., 23 (2011) 4248-4253.

34. S.J. Kim, H.J. Koh, C.E. Ren, O. Kwon, K. Maleski, S.Y. Cho, B. Anasori, C.K. Kim, Y.K. Choi, J. Kim, Y. Gogotsi, H.T. Jung, Metallic $Ti_3C_2T_x$ MXene gas sensors with ultrahigh signal-to-noise ratio, ACS Nano, 12 (2018) 986-993.
35. S.H. Hosseini-Shokouh, J. Zhou, E. Berger, Z.P. Lv, X. Hong, V. Virtanen, K. Kordas, H.P. Komsa, Highly selective H_2S gas sensor based on $Ti_3C_2T_x$ MXene-organic composites, ACS Appl. Mater. Interfaces, 15 (2023) 7063-7073.
36. T. He, W. Liu, T. Lv, M. Ma, Z. Liu, A. Vasiliev, X. Li, MXene/SnO_2 heterojunction based chemical gas sensors, Sensors Actuators, B Chem., 329 (2021) 129275.
37. S. Liu, M. Wang, G. Liu, N. Wan, C. Ge, S. Hussain, H. Meng, M. Wang, G. Qiao, Enhanced NO_2 gas-sensing performance of 2D Ti_3C_2/TiO_2 nanocomposites by in-situ formation of Schottky barrier, Appl. Surf. Sci., 567 (2021) 150747.
38. Y. Wang, H. Tan, L. Sang, Y. Xie, F. Jia, T. Zhou, G. Yin, F. Sun, X. Wang, Enhanced acetone sensing performance of Ti_3C_2 MXene/α-Fe_2O_3 nanorod composite, J. Mater. Sci., 58 (2023) 5319-5333.
39. N. Li, Y. Jiang, C. Zhou, Y. Xiao, B. Meng, Z. Wang, D. Huang, C. Xing, Z. Peng, High-performance humidity sensor based on urchin-like composite of Ti_3C_2 MXene-derived TiO_2 nanowires, ACS Appl. Mater. Interfaces, 11 (2019) 38116-38125.
40. X. Wang, D. Zhang, H. Zhang, L. Gong, Y. Yang, W. Zhao, S. Yu, Y. Yin, D. Sun, In situ polymerized polyaniline/MXene (V_2C) as building blocks of supercapacitor and ammonia sensor self-powered by electromagnetic-triboelectric hybrid generator, Nano Energy, 88 (2021) 106242.
41. S.M. Majhi, A. Ali, Y.E. Greish, H.F. El-Maghraby, S.T. Mahmoud, V_2CT_x MXene-based hybrid sensor with high selectivity and ppb-level detection for acetone at room temperature, Sci. Rep., 13 (2023) 3114.
42. B. Liu, W. Xie, H. Li, Y. Wang, D. Cai, D. Wang, L. Wang, Y. Liu, Q. Li, T. Wang, Surrounding sensitive electronic properties of Bi_2Te_3 Nanoplates – Potential sensing applications of topological insulators, Sci. Rep., 4 (2014) 4639.
43. M. Mousavi, M.B. Ghasemian, J. Han, Y. Wang, R. Abbasi, J. Yang, J. Tang, S.A. Idrus-Saidi, X. Guan, M.J. Christoe, S. Merhebi, C. Zhang, J. Tang, R. Jalili, T. Daeneke, T. Wu, K. Kalantar-Zadeh, M. Mayyas, Bismuth telluride topological insulator synthesized using liquid metal alloys: Test of NO_2 selective sensing, Appl. Mater. Today, 22 (2021) 100954.
44. B. Das, S. Riyajuddin, K. Ghosh, R. Ghosh, Room-temperature ammonia detection using layered Bi_2Se_3/Bi_2O_3: A next-generation sensor, ACS Appl. Electron. Mater., 5 (2023) 948-956.
45. Y.L. Hong, Z. Liu, L. Wang, T. Zhou, W. Ma, C. Xu, S. Feng, L. Chen, M.L. Chen, D.M. Sun, X.Q. Chen, H.M. Cheng, W. Ren, Chemical vapor deposition of layered two-dimensional $MoSi_2N_4$ materials, Science, 369 (2020) 670-674.
46. J. Dong, R. Huang, D. Hou, J. Tan, C. Song, W. Zhang, Pd, Ag decorated $MoSi_2N_4$ monolayer: A potential material for reusable CO and NO gas-sensitive material with high sensitivity, Mater. Today Commun., 37 (2023) 107354.
47. Y. Linghu, T. Tong, C. Wu, Cu-doped $MoSi_2N_4$ monolayer as a potential NH_3 sensor, Chemphyschem, 24 (2023) e202200712.
48. A. Majed, M. Kothakonda, F. Wang, Eric N. Tseng, Kaitlyn Prenger, Xiaodong Zhang, Per O.Å. Persson, Jiang Wei, Jianwei Sun, Michael Naguib, Transition metal carbo-chalcogenide "TMCC": A new family of 2D materials, Adv. Mater., 34 (2022) 2200574.

Index